CODELESS DATA STRUCTURES AND ALGORITHMS

LEARN DSA WITHOUT WRITING A SINGLE LINE OF CODE

Armstrong Subero

Apress®

Codeless Data Structures and Algorithms: Learn DSA Without Writing a Single Line of Code

Armstrong Subero
Basse Terre, Moruga, Trinidad and Tobago

ISBN-13 (pbk): 978-1-4842-5724-1 ISBN-13 (electronic): 978-1-4842-5725-8
https://doi.org/10.1007/978-1-4842-5725-8

Managing Director, Apress Media LLC: Welmoed Spahr
Acquisitions Editor: Shiva Ramachandran
Development Editor: Rita Fernando
Coordinating Editor: Rita Fernando

Cover designed by eStudioCalamar

Distributed to the book trade worldwide by Springer Science+Business Media New York, 1 New York Plaza, New York, NY 100043. Phone 1-800-SPRINGER, fax (201) 348-4505, e-mail orders-ny@springer-sbm.com, or visit www.springeronline.com. Apress Media, LLC is a California LLC and the sole member (owner) is Springer Science + Business Media Finance Inc (SSBM Finance Inc). SSBM Finance Inc is a **Delaware** corporation.

For information on translations, please e-mail rights@apress.com, or visit www.apress.com/rights-permissions.

Apress titles may be purchased in bulk for academic, corporate, or promotional use. eBook versions and licenses are also available for most titles. For more information, reference our Print and eBook Bulk Sales web page at www.apress.com/bulk-sales.

Any source code or other supplementary material referenced by the author in this book is available to readers on GitHub via the book's product page, located at www.apress.com/9781484257241. For more detailed information, please visit www.apress.com/source-code.

Printed on acid-free paper

To all those who tinker to keep the world running.

Contents

About the Author

Armstrong Subero started learning electronics at the age of 8. From then on he got into programming and embedded systems development. Within the realm of programming, he has a special interest in algorithms and data structures and enjoys implementing them in different programming languages and on different processor architectures, particularly resource-constrained systems. He currently works at the Ministry of National Security in his country, and he has degrees in computer science and liberal arts and sciences from Thomas Edison State University. He is the author of *Programming PIC Microcontrollers with XC8* (Apress, 2018).

Acknowledgments

I want to thank my family.

I want to thank *everyone* who ever said anything positive to me or taught me something. I heard it all, and it meant something.

I want to thank God most of all, because without God I wouldn't be able to do any of this.

Introduction

There is one question beginners and even good self-taught developers ask me all the time: "How do I learn data structures and algorithms?" I have had to answer this question so often that I thought I would write it down in a book so that anyone interested in this topic could follow along and understand the process.

The thing is, anyone can learn the basics of data structures and algorithms. There really isn't much to them. The hardest part of teaching someone this topic is that they are likely to be using any language from the current "zoo" of languages. They may be using Python, Java, C++, or even C or Rust. Whatever the language they are using, it is important to be able to understand data structures and algorithms at the fundamental level. As such, I decided to make my book "codeless." This is a book on algorithms and data structures that doesn't use a single line of code from any programming language.

As someone who must regularly change between different programing languages, take it from me, once you understand the concepts behind these algorithms in a codeless way, then you will be able to apply them to whatever language you are using.

There are also some people "devout" in their programming language who are unwilling to look at any material that isn't in their language of choice. As such, I have written this book without offending anyone's "beliefs." I think there are enough books on "data structures and algorithms in X language" that are thousands of pages detailing programs and all their nuances for the language you are using. I think such books would be fine complements to this one, as this book will give you the "big picture," and then you can look at whatever book you want for the rest. You can learn about a concept here and then go to that book to learn the programming details.

Everything is explained in plain English, and each chapter is short and to the point. You will learn a lot without burning your brain out.

Who Is This Book For?

This book is for people who want to understand data structures and algorithms but don't want unnecessary details about quirks of a language and don't have time to sit and read a massive tome on the subject. This book is for people who want to understand the concepts of algorithms and data structures in plain English. I assume, though, that you have knowledge of at least one programming language, preferably a C-based one such as C, C++, Java, C#, or Python. The types and terminology used in this book are biased toward people who have used one of these languages.

I assume you are good at thinking abstractly and at least have basic knowledge of a programming language and of the basics of computer science (what a bit is, what a byte is, etc.). I also assume you know basic mathematics, at least algebra, though this book is by no means a "math-heavy" book. I use math concepts only where they are necessary and explain them clearly.

What Will I Need for This Book?

You won't need anything except an open mind and time to read and understand the concepts being presented. I wrote this book as if we were having a coffee and I was explaining these concepts to you from the top of my head. For that reason, you won't need any compiler or text editor, just a listening ear.

What Will I Learn in This Book?

You will learn quite a lot in a short space of time about data structures and algorithms. You will learn the concepts of data structures and algorithms with a focus on the most relevant ones in simple, plain language. After completing this book, you'll understand which data structures and algorithms can be used to solve which problems and will be able to confidently follow along with discussions involving algorithms and data structures.

PART I – Data Structures

Chapter 1 goes into what algorithms and data structures are, and we discuss primitive types and Big O notation.

Chapter 2 looks at the linear data structures of arrays and linked lists; we also discuss stacks and queues.

Chapter 3 delves into trees and tree-based data structures.

Chapter 4 introduces you to hash-type data structures.

Chapter 5 briefly covers the basics of graphs.

PART II – Algorithms

Chapter 6 introduces two common algorithms, that of linear search and binary search.

Chapter 7 explores sorting algorithms, including bubble sort, selection sort, insertion sort, merge sort, and quick sort.

Chapter 8 presents some search algorithms; we look at breath-first search, Dijkstra's algorithm, and the A* algorithm.

Chapter 9 introduces clustering algorithms, specifically the K-means algorithm and K-nearest neighbor and get a taste of machine learning and neural networks.

Chapter 10 teaches some basics about the concept of randomness.

Appendix A provides resources for going further.

Upon finishing this book, you will have a solid foundation in algorithms and data structures and will be able to use this knowledge when writing and designing your own programs.

Errata and Suggestions

If you find any errors in this text, have suggestions, or want to ask a question about your own project, you can contact me at armstrongsubero@gmail.com; please, no spam.

Data Structures

Intro to DSA, Types, and Big O

Every journey has a beginning. In this chapter, we will begin our journey by talking about what data structures and algorithms are. I will introduce basic types, and I will show you how easy Big O notation is to understand. If you have ever read a dull, overly complex book on data structures and algorithms (abbreviated to DSA), you will love how quickly this chapter teaches you things!

An Overview of Data Structures and Algorithms

"Talk is cheap. Show me the code."

—Linus Torvalds, Finnish software engineer and creator of Linux

Linus said this while replying on the Linux Kernel mailing list on August 25, 2000. This reply has become a famous quote among programmers. It is used by developers whenever they don't want to read anything and instead just jump into coding. This approach is particularly taken by novices and poorly self-taught programmers. They don't plan, they think they know more than everyone, and they think programming is all about the code. This couldn't be further from the truth. Code simply expresses your thoughts to solve a problem.

© Armstrong Subero 2020
A. Subero, *Codeless Data Structures and Algorithms*,
https://doi.org/10.1007/978-1-4842-5725-8_1

Nothing more. Therefore, the more you know, the more you can apply to solve a problem.

Data structures and algorithms are simply more things to know to apply to solve your problems. Despite some people using them interchangeably, data structures and algorithms are actually very different things. It is possible to learn data structures and algorithms without touching a programming language. Programming essentially consists of thinking algorithmically and knowing the syntax of the programming language to solve problems. In this book, we will focus on thinking algorithmically and avoid learning the syntax of any programming language.

Before we discuss data structures and algorithms, I think we should talk a little about data. Data can mean different things depending on the discipline you are currently occupied with. However, *data* within the context of this book refers to any information that is stored on your machine or that is being handled or processed by it. Data should not be confused with *information*, which is data that has been processed; however, within the context of computing, many developers may use these terms independently to mean the same thing.

Data Structures

A *data structure* is a concept we use to describe ways to organize and store types of data. Data structures are important because they not only provide a way of storing and organizing data but also provide a way of identifying data within the structure; additionally, they show the relationships of the data within the structure. It is best to illustrate what a data structure is with an example.

For example, let's say we have some footwear, as depicted in Figure 1-1. We have two boots and two shoes arranged alternately.

Figure 1-1. Two types of footwear

We can think of each side of the shoe as a unit of data. If we needed a way to maintain this data arrangement, we would need a mechanism to provide some ordering and identification of these shoes; this is what we may call a *data structure*. A data structure may provide some mechanism to organize and store this data, for example, by separating boots and shoes as shown in Figure 1-2.

Figure 1-2. Separating boots and shoes

A data structure may also be able to take each item and, regardless of whether it's a shoe or boot, assign an identifier, for example, a number as shown in Figure 1-3.

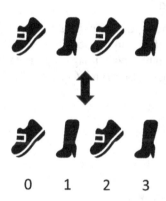

0   1   2   3

Figure 1-3. Assigning identifiers

So, if I wanted the "second shoe," instead of wondering if this means from the left or from the right, I can simply tell the data structure, "Give me the shoe at index 2," and I will get exactly what I wanted.

As basic as it may seem, this is all a data structure does, though there are much more complex methods of storing, identifying, and showing the relationships within data. This explanation provides a good grasp of what a data structure does. If you still aren't clear what a data structure is, it is best to think of a data structure as a container for data.

Algorithms

An *algorithm* is a method of solving a problem by applying a sequence of steps that will always work to solve the type of problem it was designed to solve. Another way of saying this is that an algorithm is simply a method of solving a problem in an ordered manner. We can even shorten it further to say an algorithm is a procedure. Many people may have expanded or diminished perspectives to describe what an algorithm is, but this definition will work for our understanding.

One thing everyone will agree on is that algorithms are logical steps to accomplish a task. To accomplish this, an algorithm must be simple, precise, and unambiguous. Though some programmers focus on using esoteric features of programming languages that make an algorithm hard to read for other programmers, the simpler and more intuitive an algorithm is, the more powerful and useful it will be.

We can describe algorithms with natural languages such as English, pseudocode, or a programming language. We can discuss algorithms at great length; however, it is best to see how algorithms work with a good example. In our example, we can show how algorithms operate with a pure English description, with no code needed.

We can gain an understanding of algorithms by looking at one in action. Let's say you need to put some fruit on a plate. You would, of course, grab a plate and put some fruit on it, right? Well, imagine you were describing these same steps to a robot. How would you do it? You would tell the robot to do something like the following:

1. Go to the cupboard.
2. Open the door.
3. Take out a plate.
4. Go to the fruit bowl.
5. Put the fruit on the plate.

This seems like a logical sequence of steps, until you look around the kitchen and realize that the robot left the cupboard door open. So, you decide to add another step to the algorithm, as shown here:

1. Go to the cupboard.
2. Open the door.
3. Take out a plate.
4. Close the cupboard door.
5. Go to the fruit bowl.
6. Put the fruit on the plate.

You watch the robot run through the new steps, only to realize that the robot is stuck at step 6, unable to put any fruit on the plate because the fruit bowl has a tomato and the robot doesn't know if a tomato is a fruit or a vegetable. This is where data structures come into play for working with algorithms.

Algorithms and Data Structures in Tandem

Data structures and algorithms are separate but complementary to each other. We can say that data structures organize data that algorithms work upon. Data structures provide the fragments that algorithms take in, process, and output into whole information that the end user wants to derive.

Let's look at the problem we arrived at in the previous section when we wanted our robot to place fruit on a plate. While our algorithm seemed sound, because of a simple classification problem, our robot became stuck. One solution to this is to tell the robot that if the item is part of a plant, it is a vegetable, and if the item is formed from the mature ovary of a seed, then it is a fruit. The robot may become confused here because that is too much information to process. We must define what a plant is, what a seed is, what an ovary is, and what a vegetable is, which makes our algorithm more complex.

Another, simpler way to do this is to have a data structure that organizes images with all the fruit the robot is likely to encounter and perform a check to see whether the item belongs to a member of that structure. If it is a member, the robot places it on the plate; otherwise, the robot skips that item. If the item is in between, then the robot will place it if it shares more characteristics with other fruits than not. This "grouping if lots of characteristics are shared" is part of a later algorithm for classification we will discuss in Chapter 9 on clustering algorithms.

Primitive Types

In the previous section, we talked about data structures and algorithms. We said that a data structure stores and organizes data that algorithms then process to get some output. We also said that data is information stored on your machine or that is being processed by it. What we didn't say is that this data can be of different types.

Data needs to be of different types because a computer is a very specific device. We have all heard the phrase "garbage in garbage out" (GIGO). This term gives us only part of the picture. We must realize that a computer is an input-output device, yes, but it is also a device that performs processing on data before it is output. This means we can put good data in and still get garbage out.

This can happen in two ways. The most obvious way is that the computer will perform an error in the processing of the problem (our algorithm is bad). The other way is that if we fail to tell the computer what type of data it is working on, then it will not know what to do and will still output garbage. This will occur even if our algorithm is good, as the algorithm is expecting data of a certain type.

This seems so simple, yet it is so important. In fact, it is so primitive that we have a name for telling the computer what type of data types we are working on. We call this the *primitive* data type.

A data type will tell our machine the attributes of the data we are using and what we intend to do with the data. There are many data types in computing, but the primitive data type is the most basic type there is in your programming language of choice. It is so basic that it usually forms part of the language itself and is usually built into the language.

From these primitive languages, we get other data types. Though most programming languages will have their own names for the data types, they all fall into one of the following from the C-based languages. I call them the "Big Four" because from these big four cornerstones, everything else is built upon.

Sometimes these data types are also called *atomic data types* simply because you cannot divide them into a lower-level type; they are as low as a data type can get.

These are the four types of primitive data types:

- Boolean
- Character
- Integer
- Floating-point number

These four basic data structures tell your computer what type of data you are working on. Let's look at each of them in detail now and get a good understanding of what they are.

Boolean

The Boolean primitive data type is the first one we will look at. The Boolean type is known as a logical data type, having two possible values: True and False, 0 and 1, or ON and OFF.

At the heart of classical computing is the Boolean data type. Classical computers have a processor that consists of millions of transistors that turn on or off. Based on them being on or off, we get all sorts of amazing operations, storage, and all the mechanisms you can think about in classical computing. The binary running in the arithmetic logic unit (ALU) and the instruction register consists of simply 0s and 1s. The concept of ON and OFF is so powerful that it translates into basically all programming languages in one form or another.

While it is not necessarily pertinent to our discussion on the Boolean data type, it is handy to know that in the realm of quantum computing there exists quantum bits called *qubits*. While in the realm of classical computing the "hard on" and "hard off" Boolean type is standard, quantum computers have qubits that are neither 0 nor 1 but are in a state of superposition having both states simultaneously.

Character

The next item on our list is the character. Classical computers work on binary bits of information and prefer to crunch numbers. Humans, on the other hand, prefer to work with the natural language of their choice. This is where characters come in.

Us humans find it difficult to remember data by its binary sequence of representation. As such, characters can be used to give us a more understandable and easier-to-remember version of this sequence of binary bits. When we combine this sequence of characters into strings, it makes this much better. For example, the variable cat is more readable, recognizable, and memorable written as *cat* than as 01100011 01100001 01110100. Characters are used to make up strings and have differing encodings depending on the machine you are using.

Integer

The computer is all about numbers, and no data type is better fit to represent numbers than the integer. An integer refers to a mathematical number that does not have any fractional constituent, for example, 6 or 7.

The only distinction between integers in mathematics and integers in computer science is that in mathematics integers, unless they belong to a finite subset, are infinite. In computing, because of limitations in processor architecture and memory, integers are not infinite.

Floating-Point Number

The floating-point number can be thought of as an integer with a fractional constituent. This is the type of number that people refer to as decimals, for example, 6.5 or 7.1. They are called *floating point* because the little dot (.) called the *radix point* can float, or move, to accurately represent the number we are displaying.

Floating-point numbers are usually described as either single precision or double precision. Single-precision floating-point numbers are represented with a 32-bit word, whereas double-precision numbers are represented by a 64-bit word. Usually single-precision floating-point numbers are called *floats*, and double-precision floating-point numbers are called *doubles*. There are also floating-point numbers that require 128-bit word representations that are called *decimals*.

Functions

Before we go further, I would like to take a moment to discuss an important mathematical concept that is intertwined with computer science, that of the function. Note that the mathematical function we are discussing shouldn't be confused with functions in programming, which refer to a section of code that performs some task.

In mathematics, a *function* is an expression that maps an independent variable to a dependent variable.

We call the input to the function the *domain*, and we call the output the *range*. Each domain element can be mapped to one, and only one, element in the range. The function determines which element at the domain gets mapped to which at the range.

In Figure 1-4 we have a function that matches numbers on the domain with letters in the range.

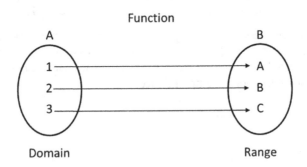

Figure 1-4. Mathematical function

The function can be visualized as a black box that takes an input, does some processing, and gives an output (kind of sounds like a computer, doesn't it?). Look at Figure 1-5 to see the essence of what a function is.

Figure 1-5. Essence of a function

Even though earlier it was mentioned that a mathematical function shouldn't be confused with a function in programming, in programming when we talk about a function, we talk about a routine or procedure. Some functions in programming are identical to the functions in mathematics in the sense that they take some input parameter, do some processing on it, and then return a value. This simple concept is important to understand, as you will see moving forward.

Functions, Methods, Procedures, and Subroutines

This book takes a codeless approach to teaching you data structures and algorithms; however, there is some lingo from programming languages that you need to be familiar with as some terms will come up in our discussion of algorithms in later chapters. These are functions, methods, procedures, and subroutines.

It is best to begin our discussion on functions by relating it to something we are familiar with like a house. When we are building a house, we use bricks to gradually build it up; some people call these bricks *blocks*. Now you can think of a program like a large house made of many building blocks. A group of blocks can be combined to make a wall or a room. According to how you configure the blocks, you can have a wall that has a window or a wall that does not have a window.

Within the realm of construction, it is not possible to save a configuration of bricks as a wall with or without a window and simply replicate it to build our house because actual people must assemble the wall. However, when building software, it is possible to save an arrangement of a block of code that solves a task and use it many times over. We call this bit of code a *function*.

This function is like the mathematical function we spoke of in the previous section; however, it is not the same. Functions in programming languages take data as an input that we call *parameters* or *arguments*, and sometimes they return a result. I say sometimes because unlike mathematical functions that give an output based on an input, functions sometimes do not return any result, which is called a *void* function in C-based programming languages.

Some programming languages are called *object-oriented programming* (OOP) languages and within them exists special code that are blueprints for other code. We call this code that acts as a blueprint a *class*. When functions exist inside of a class, we call them *methods*. Essentially, methods are functions called by that name in discussions about OOP languages.

Some programming languages may call a function a *procedure* or a *subroutine*; you should know, though, that some programmers refer to functions that do not return a value as procedures as well.

Functions, methods, procedures, and subroutines all mean the same thing and serve the same purpose; it depends on the language you are using and sometimes who you are speaking to. They are all small programs, or subprograms rather, that can be called within a larger program.

Recursion and Iteration

Before we go further with our discussion on algorithms, I think we should look at the concepts of recursion and iteration. Recursion is a tricky thing to understand. Let's imagine you hear a new word such as *thingamaword* and you don't know what it means. You turn to the dictionary only to see the following definition:

> **thingamaword. see thingamaword.**

Such a definition would make no sense, as the meaning refers to itself. This is what *recursion* is; it is the process of something being defined in terms of itself. Within the context of programming, a recursive function is one that calls itself as part of its operation. To end the cycle of a function calling itself, usually a recursive function runs until a certain condition is met.

If the recursive function calls itself too much, then due to a machine having finite memory, the machine will run out of memory, and you may get what is known as a *stack overflow error* due to the maximum recursion depth being exceeded.

Recursion finds many uses in programming languages, and many algorithms are in fact dependent on recursion to operate. We will see examples of such algorithms as we work throughout the book.

Iteration is another term that you hear quite often that should not be confused with recursion. *Iteration* within algorithms refers to the process whereby a block of code sequence is repeated until a certain condition specified by the program designer is met.

For example, let's say we are entering a password into a system. The program can be designed in such a way that the user will enter the password a fixed number of times until the correct password is entered. An iterative function can be the heart of the program in that the user will be given three times to enter the password, and if the three times are exceeded, then the user will be locked out of the program.

In such a scenario, the iteration is the block of code being executed three times, and the specified conditions for exiting the iterative sequence are the correct password being entered within three tries and the user entering the wrong password.

Iteration finds extensive use in the execution of algorithms and within general programming as well. One thing that should be mentioned about iteration, though, is that, like recursion, if you do not specify an exit condition due to a computer having finite memory, you will get errors in the program.

Many algorithms can use either recursion or iteration to solve the problems they are designed for; however, it depends on the algorithm designer to choose which method is best.

The Three Types of Algorithms

In ending our introduction on data structures and algorithms, I think it's best if we discuss the three types of algorithms. Within the context of algorithm types there are three basic approaches to solving a problem. These three approaches are dividing and conquer, greedy, and dynamic.

Divide-and-conquer algorithms take a large problem, divide it into many smaller problems, and then combine the results to obtain a solution. Think about ants harvesting leaves from a tree. If the goal of the ants is to strip the tree of all its leaves, this is a daunting task for any single ant regardless of how hard it works.

A tree is many times larger than an ant; however, if thousands of ants each attack a single leaf, then they would be able to solve the problem of stripping the tree of its leaves. Some large problems within computing require this type of approach to solve a problem.

Greedy algorithms are algorithms that take the best decision at any point in time. Imagine a superhero having a choice to make between catching the bad guy who is escaping or rescuing the damsel who is about to face impending doom. This requires decisive action, and the superhero will make the choice he thinks is best at that point in time.

Similarly, greedy algorithms will do what is best at the point in time at which the algorithm is executing, whether it has the best overall impact to solve the problem at hand or not. Some problems require this type of algorithm to solve. One of the most popular is the traveling salesperson problem. This problem is one of the most important in computer science.

It goes like this: a salesperson wants to visit X number of cities. What is the shortest path he must take to travel through all the cities while traveling the minimum possible distance? While the solution plagues computer science, greedy algorithms offer somewhat of a solution to the problem. If we take a greedy approach, we will begin at one city and take the next closest possible city thereafter for each city we visit. While this may not be the shortest overall path, it may give a good approximation of what the best possible path is.

In contrast to the greedy approach, dynamic programming–based algorithms take another approach to solving problems. With the dynamic approach, we make decisions catering for future implications while considering the past results. While both approaches focus on breaking up a problem into smaller parts, the greedy algorithms consider only what is best at that point in time; the dynamic approach looks at multiple solutions to the problem, computes them, stores them, and then later will recall them for reuse.

The dynamic approach is used for things such as speech recognition, gene sequencing, and, probably the most likely use case you will encounter, matrix chain multiplication. The best way to describe the dynamic approach is that while the greedy approach approximates, the dynamic approach optimizes.

Analyzing Algorithms

We haven't even gone into looking at any algorithms yet, but I think it's time we discuss some basic algorithmic analysis as it will make everything going forward much easier.

Designing an algorithm is one thing. The other thing we must consider is analyzing the performance of the algorithm once we have designed it. As we discussed, an algorithm is a sequence of steps that is used to solve a problem. The fewer steps we can use to solve our problem, the more efficient our algorithm can be said to be. There are two ways we can analyze our algorithms for efficiency. They are called *time complexity* and *space complexity*.

Time complexity refers to the amount of time an algorithm takes to solve the problem based on the input it is given. This is the most common method of examining how efficient an algorithm is in terms of performance.

Space complexity, on the other hand, refers to the amount of memory from the machine the algorithm will take to solve a problem. This method is not as widely used for explaining algorithm efficiency and analysis. This does not mean it is not necessary as there are instances where space complexity is important, particularly when working on resource-constrained systems; it is just not as widely used.

That's all there is to it. Though both methods of analyzing algorithms are important, as was mentioned, time complexity is by far more often used in practice for analyzing algorithms and measuring their performance, and for that reason we will focus on time complexity in this chapter.

We will need to delve a little deeper into time complexity to understand the algorithms presented in this book more effectively. There are two ways we can analyze algorithms' time complexity, which are practically and mathematically.

Practically, we can write the algorithm in our programming language of choice and vary the input and output and then use some method to record our observations based on the amount of input and/or output data our algorithm must work with. Based on this observation and recording, we can then determine how efficient our algorithm is. Not only is this method dreary, but it is inaccurate and limited in scope. To circumvent this way of doing things, we can use a mathematical method.

The way we mathematically determine the efficiency of an algorithm is known as *asymptotic analysis*. While it seems complex and unnecessary, using asymptotic analysis will save us a lot of time when analyzing our algorithms.

Essentially, asymptotic analysis describes the limiting behavior of an algorithm; we usually try to determine what would be the worst-case performance of our algorithm, and sometimes we try to find the average performance. To effectively describe the differences in asymptotic growth rates of functions, we use something called Big O notation, which we will look at in the next section.

Big O

The way we mathematically determine the efficiency of an algorithm can fall into one of three categories. We can measure the worst-case scenario, which is the longest time our algorithm will take to run; we can measure the best-case scenario, which is the shortest time our algorithm will take to run; or we can measure both cases, which is both the best case and the worst case. Sometimes we can also measure the average case, which is how we expect our algorithm to run most of the time.

There are many notations we use for describing the efficiency of our algorithm including Big Omega, Little Omega, Big O, Little O, and Theta. Each notation has its application; however, some methods are used in the industry more frequently than others.

Of all these methods, Big O is the one that is most utilized for asymptotic analysis notation. If you have heard of this notation, then you may wonder why it is called Big O. Well, the *O* in Big O refers to the "order of complexity" or the "order." We usually write the order of complexity of something with a capital *O* or a "big *O*." That is why we call it Big O notation.

In other forms of notation, omega notation describes the measurement of the minimum time our algorithm will take to run, and theta describes the minimum and maximum time our algorithm will take to run. By contrast, Big O describes the maximum amount of time our algorithm will take to run. In essence, Big O describes the worst-case running time of our algorithm.

There are many ways to describe Big O runtime, but here are some types of Big O run times you will come across:

- **O(1)**: The algorithm has a constant execution time, which is independent of the input size.

- **O(n)**: The algorithm is linear, and performance grows in proportion to the size of the input data.

- **O(log n)**: The algorithm is logarithmic; as time increases linearly, then n will go up exponentially. This means that more elements will take less time.

- **O(n log n)**: The algorithm is logarithmic multiplied by some variable n.

- **O(n^2)**: The algorithm is quadratic, and the running time is proportional to the square of the input.

- **O(2^n)**: The algorithm is exponential; execution time doubles with the addition of each new element.

- **O(n!)**: The algorithm is factorial and will execute in n factorial time for every new operation that is performed.

These classifications are the ones you will come across most often. Constant time has the best performance followed by logarithmic time. The factorial O(n!) has the worst time possible; it is terrible! In fact, it is at the heart of a complex computer science problem, the traveling salesperson problem that we discussed earlier.

Conclusion

In this chapter, we got an introduction to data structures and algorithms. We discussed data, functions, data structures, and algorithms. We also looked at primitive data types, iteration, and recursion as well as the three fundamental types of algorithms. We then went a step further and discussed analyzing algorithms as well as the basics of Big O notation. This was a lot of information, and believe it or not there are many practicing developers who do not understand some of the topics presented in this chapter, so you are already ahead of the pack. In the next chapter, we will begin our "deep dive" into the world of DSA, as we will discuss linear data structures.

Linear Data Structures

In the previous chapter, we looked at an overview of data structures and algorithms. From this chapter forward, we will explore the types of data structures there are, beginning with the simple linear data structures. It is important to understand linear data structures because they get you into the mind-set of what a data structure is rather intuitively. If you have experience with data structures, you'll like how simply the topic is presented. If this is your first time learning data structures, you'll marvel at how simple they are to understand. This is the introduction I wish I had to the topic. Let's get started with linear data structures!

Computer Memory

Before we delve into linear data structures, it is important to understand a little about computer memory. While it may seem unnecessary, remember that to understand data structures, we must understand a little of the underlying hardware, particularly memory organization.

Computer memory is to data structures as water is to a fish: it is essential for existence. As you will come to realize, the aim of many data structures is to effectively use the available resources, and one such resource is memory. Once we have a basic understanding of computer memory, we will be able to understand how linear data structures operate.

© Armstrong Subero 2020
A. Subero, *Codeless Data Structures and Algorithms*,
https://doi.org/10.1007/978-1-4842-5725-8_2

When we are talking about computer memory, we are talking about a type of workspace where the computer can store data that it is working on or has worked on. Essentially, memory is a storage space for the computer.

As a side note, something that we should know about memory is the terminology given to the binary prefix used to describe computer bits and bytes. We are all familiar with the terms *kilobytes*, *megabytes*, *gigabytes*, and *terabytes* that describe computer memory in terms of the metric system. Let's take kilo, for example. In the metric system, *kilo* refers to 1,000; however, in the world of computing, when we say a *kilobyte*, we are usually referring to 1,024 bytes.

So, while in the usual context *kilo* means 1,000, in computing *kilo* would mean 1,024 since the International System of Units (SI) metric system did not have a way to represent units of digital information. Now, this is inconsistent, and within computer science, like all sciences, consistency in key. Thus, we refer to the units used in the SI system as the *decimal prefix* for describing measurement of memory.

To add some consistency, the International Electrotechnical Commission (IEC) units are now used when describing computer memory in some instances. The IEC units use a *binary prefix* for describing memory as opposed to a decimal prefix. So, instead of *byte* attached to the prefix there is now binary attached to the prefix.

This means that *kilobyte* (KB), *megabyte* (MB), and *gigabyte* (GB), for example, are now replaced with *kilobinary* (Ki), *megabinary* (Mi), and *gigabinary* (Gi). They are also referred to as *kibibyte*, *mebibyte*, and *gibibyte* as a contraction of their binary byte names. I use the IEC standards in this book when referring to memory size.

Computer memory is hierarchical, and this is important as each memory type on the machine is specialized. When we say something is hierarchical, we mean there is some order based on rank. In everyday life there are many hierarchical structures, and within computer memory there is no difference. Groups like the military and many systems in the natural world depend on a hierarchical structure for success.

Though there are many ways you can represent a hierarchical structure, a pyramid-based, stack-like structure is ideal for representing computer memory. A good way to visualize memory arrangement within a computing device is like the memory stack in Figure 2-1 where you have the largest memory on the bottom and the smaller memory on the top of the stack.

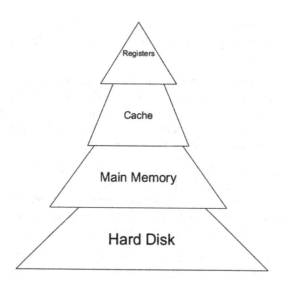

Figure 2-1. Memory stack

This arrangement does not mean that any memory is more important than another as each component is necessary to make up the entirety of computer memory. Instead, this stack can be thought of as a sort of size hierarchy with the typical arrangement of the amount of memory that is typically present in a computing device.

If we look at our memory stack in Figure 2-1, we will see at the base of the structure is a hard disk. In computing devices, there is a storage device, which is typically a solid-state drive (SDD) or a hard disk drive (HDD), that we call *disk storage* or the *hard disk*. The hard disk is the component that contains the data that is loaded into main memory called *random access memory* (RAM).

The next layer of the stack is the RAM. RAM is called the main memory of the computer and is indeed the main memory of modern-day computing devices. RAM is further loaded into a type of memory called *cache* that resides on the CPU chip. The information in cache is eventually loaded into the registers that contain data the CPU is presently working on.

Let's say you are loading a word processing software. The software would reside on your hard disk memory. When you launch the program, it is transferred from the hard disk into RAM.

As you move up the stack from RAM, the next level you encounter is the cache memory. Within the cache there are generally two different caches; there is the LI cache, which is fast and is almost as fast as the CPU registers (more on this later). There is also the L2 cache, which is slower than LI but faster than RAM. Cache memory stores information that is most likely to be used by the CPU, making some operations extremely fast.

Cache is small, not exceeding a few mebibytes, and generally is in the kibibyte range. L1 cache is typically a few tens to just over 100 KiB, and L2 cache is a few hundred kibibytes.

Some CPU manufacturers also have an L3 cache, which is a few mebibytes. While cache has many benefits, it becomes inefficient if it is too large. Part of the reason cache speeds up operations is that it is much smaller than RAM, so we locate information in the cache easily. However, if the cache is too large, it becomes like RAM and the speed benefits decrease.

The example many people use is that of a textbook and your book of notes. When you have a textbook, the 1,000+ pages of information can be summarized into useful notes you are most likely to use or information that is important to you. Should you rewrite all 1,000+ pages of information in your notebook, then the efficiency of the textbook will diminish.

At the top of the stack is the register. The register is as fast and as small as memory can get within a computing device. Registers are memory locations that are directly built into the processor. They are blazingly fast, and low-level languages can directly access this type of memory.

Memory can also be arranged into a hierarchy by speed. In Figure 2-2 we see this memory speed hierarchy. Within this structure we see that disk memory is the slowest, and register memory is the fastest.

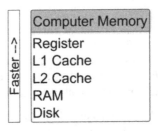

Figure 2-2. Memory speed hierarchy

Within computer memory you often hear people talking about the memory locations called the *memory address*. The memory locations are known as physical addresses, whereas the operating system can give programs the illusion of having more memory by providing memory mappings to these physical addresses. We call these mapped representations of actual physical locations *virtual memory*. What all this means is that though the actual memory locations are finite and thus can be exhausted, the operating system can provide proper management of memory to ensure programs using the memory have enough to run properly.

As we said earlier, virtual memory is a representation of physical memory space. Virtual memory contains virtual addresses that specify the location in virtual memory our program is accessing. This virtual memory is divided into virtual pages, which are used by a page table that performs address translation of virtual page map addresses to physical addresses.

The addresses provided by the operating system to your program identify each of these virtual memory slots. When you ask the computer to store something, the operating system will check to see which slots are available and let you store it there. If you have no slots available, then you will get an error and your program will not run.

While this information may not seem to be useful, keep this arrangement in mind as we move forward with our discussion on linear data structures.

An Overview of Linear Data Structures

When we say that a data structure is linear, we mean the elements of which the data structure are comprised are arranged in such a manner that they are ordered sequentially, with the relationship between elements being an adjacent connection. Such structures are simple to understand and use when developing programs.

In this section, we will examine the most common of these linear data structures: the array and the list. These two linear structures are the two "workhorse" data structures because almost all other data structures stem from them or use them in some way.

Once you understand the array and list, you will have a solid foundation to begin working with other data structures. While they may have different names in other programming languages, the purpose and arrangement will be similar and will serve you well no matter which language you are using.

The Array

The array is one of the most fundamental data structures we have for storing and organizing data. It is such a basic data structure that it usually is part of the underlying programming language. All the popular C-based programming languages utilize arrays, so you are sure to encounter them. As simple as arrays are, they are powerful and the best starting point for any discussion on data structures.

Arrays store elements that are of the same data type. In Chapter 1, we discussed the different data types, and should you need a refresher, you can refer to that chapter. We call each of these data types an *element*, and the elements are numbered starting from 0.

The number assigned to an element is called the *index* of the array. The array has a structure like Figure 2-3, with elements arranged sequentially. These elements usually have commas separating them. This is how arrays are represented in many of the C-based programming languages, and this is the convention we will use in this book.

$$[0, 1, 2, 3, 4, 5, 6, 7, 8, 9]$$

Figure 2-3. The array structure

If we think about the arrangement of elements within the array, we will see that the elements within the array are arranged sequentially or consecutively. Because of this sequential nature, data in an array can be read randomly. However, adding and deleting data from an array can take a lot of time due to this sequential organization of elements. This is one of the limitations of the array as a data structure. There are others, as you will uncover as we go along; however, this is a prominent one.

In most programming languages, when the computer is assigning memory for an array, the size of the array must be specified by the program designer before memory is allocated for the array. This means the programming language lets you reserve the size of the array before the program is run. Other programming languages provide mechanisms by which you do not need to explicitly reserve any memory, which saves memory space and leaves less burden on the programmer.

The type of arrays we described previously is called a *one-dimensional* array. We call it a one-dimensional array because when you access elements from within the array, it is a single element you are accessing at each index, representing either a single row or a single column.

An array may also be multidimensional. A *multidimensional* array can be thought of as an array of arrays. Such structures are useful when you want to define something as a grid, since two-dimensional arrays are essentially a grid of rows and columns that you can access programmatically. The most commonly used multidimensional array is the two-dimensional array. When we store data in a two-dimensional structure, it is known as a *matrix*.

Lists

The list can be thought of as a special type of array. Whereas in an array data is stored in memory sequentially, in the list data is stored disjointed in memory. This allows for a linked list to utilize memory more effectively.

In a list, the data is organized in such a way that each data element is paired with a pointer or reference that indicates the next memory location for the next element within the list. To access a data element, we must access it through the pointer that precedes it.

The pair of a data element and a pointer to the next item in the list is known as a *node*. The list is structured in that there is an entry point into the linked list. This entry point is called the *head* of the list.

This type of linked list structure with one node having a pointer to the other is known as a *singly linked* list. In the singly linked list, the last node doesn't point to any other node and has a null value. Figure 2-4 gives a good representation of the linked list data structure.

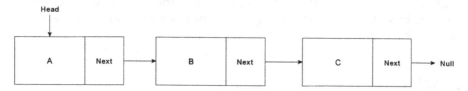

Figure 2-4. The linked list data structure

There is also a linked list structure where each node has a pointer not only to the next node in the list but also to the previous node in the list. This layout is called the *doubly linked list*. Figure 2-5 shows the doubly linked list. This layout makes the linked list more efficient when deleting data and for allowing bidirectional traversal of the list.

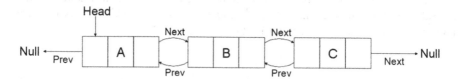

Figure 2-5. The doubly linked list data structure

Lists may also be of the type called a *circular linked list*. In a circular linked list, all the nodes are connected to form a circular arrangement. The last node is connected to the first node, and there is no null at the end of the list.

The internode connections within a circular linked list may be either singly linked or doubly linked. Figure 2-6 shows us what a circular singly linked list looks like.

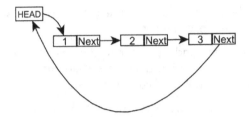

Figure 2-6. Circular singly linked list

These circular linked lists have many uses within computing, particularly uses related to buffering and other creative uses the programmer comes up with. Circular lists can also be used for queues, which is something we will look at later in this chapter.

Stacks

Stacks are a type of linear data structure where the data that is added is put into the lowest available memory location. Stacks are abstract in that we think of how they operate, but their implementation will vary based on the technology used to implement the stack and on the type of stack we are trying to construct.

The act of adding data to the stack is called a *push*. When we remove data from the stack, we do so by removing the element that was last added; this is called a *pop*. Because of this arrangement of removing data first that was last added to the stack, we call the stack a *last in first out* (LIFO) data structure. Figure 2-7 shows us the arrangement of the stack.

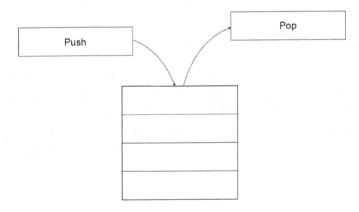

Figure 2-7. The stack

While this concept is simple to visualize, one of the key weaknesses of the stack as a data structure is in its simplicity. The stack can have data elements removed and added only from the top of the stack. This limits the speed at which a certain element can be retrieved from the stack. However, in applications such as reversing a string or backtracking, for example, the stack can be useful.

We can create stacks that are *static*, meaning they have a fixed capacity, and we can create stacks that are *dynamic*, meaning the capacity can increase at runtime. Dynamic stacks vary not only in size but also in the amount of memory they consume.

To create a static stack, we can use an array. We can also design a stack that is dynamic by using a singly linked list with a reference to the top element, which is why earlier I mentioned that while the operation of the stack will remain the same, the implementation will vary. Stacks find uses in many fundamental computing processes including function calls, scheduling, and interrupt mechanisms.

Queues

A queue is a type of data structure that assigns a priority to each element. Elements that are added first to the queue are placed at the end of the queue. The act of adding data to a queue is known as *enqueue*. When we are removing a data element from the queue, the element that has been in the queue the longest is removed first. When we extract data from a queue, we call it a *dequeue*.

Because of the queue removing data that was added to the queue first, we say that the queue is a *first in first out* (FIFO) data structure. This can be contrasted with the *last in first out* (LIFO) nature of stacks.

By combining the LIFO structure of stacks with the FIFO structure of queues, we can create some pretty amazing programs and structures, as we will see later in the book.

The queue is unique in that although the basic queue as we discussed here is a linear data structure, there are some queues that can take a nonlinear form, as we will discuss later.

In Figure 2-8 we see the FIFO method in action. New items are added to the rear of the queue for the enqueue, and requested items are removed from the front of the queue for the dequeue.

Figure 2-8. Queue

Priority Queues

A priority queue is an extension of the regular queue we discussed earlier. The priority queue, however, uses a key-value system to arrange the items within the queue. In the priority queue, every item in the queue has a priority associated with it, which can be thought of as a key. An element within the priority queue with a higher priority will be dequeued before an element with a lower priority. If we have two elements within the priority queue that have the same priority within the queue, they will be dequeued according to their order within the queue.

Sometimes items within a priority queue may have the same priority; when this occurs, then the order of the items within the queue is considered. When using priority queues, these data structures usually have methods to add items to the queue, delete items from the queue, get items from the queue with the highest priority, and check if the queue is full, among other things.

These queues can be constructed using the other data structures we examined such as linked lists and arrays. The fundamental data structures we use to construct the queue can affect the properties of the queue.

Priority queues enqueue items to the rear of the queue and dequeue items from the front of the queue. Figure 2-9 shows the priority queue.

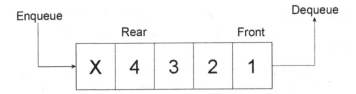

Figure 2-9. Priority queue

As we will see later, some algorithms will utilize priority queues, which have many uses in data compression, networking, and countless other computer science applications.

Conclusion

In this chapter, we looked at various linear data structures. We discussed the basics of computer memory, arrays, linked lists, stacks, and queues. While this introduction was compact, we covered a lot of information without dwelling too much on unnecessary implementation details. These basic data structures have many applications that we will see moving forward. In the next chapter, we will look at trees, which will advance your understanding of data structures.

Tree Data Structures

In this chapter, we will look at a powerful data structure, the tree. The tree is arguably one of the most utilized complex data structure types. Once you understand trees, you will be able to grasp other data structures and algorithms that will be presented throughout the book. In this chapter, we will explore the basics of trees and look at some of the more common tree data structures.

Trees

We will begin our discussion of tree-based data structures by understanding what trees are. Think about a tree. When we think about a tree, we think about a giant plant that has roots with a trunk, branches, and leaves, as shown in Figure 3-1.

© Armstrong Subero 2020
A. Subero, *Codeless Data Structures and Algorithms*,
https://doi.org/10.1007/978-1-4842-5725-8_3

Figure 3-1. Depiction of a biological tree (Image source: Freepik.com)

In computing, trees are like the ones all around our planet. Whereas the data structures we looked at previously organized data linearly, a tree organizes data in a hierarchy. Figure 3-2 illustrates this concept and gives you a visual understanding of what a computer science tree generally looks like. In fact, if you look at a computer science tree, it looks like a biological tree with the root at the top!

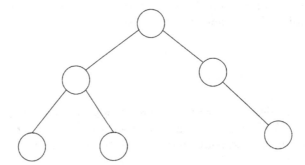

Figure 3-2. Computer science tree

In biological trees, there is a root that is responsible for keeping the plant in the ground. We can think of the root as the base responsible for keeping the plant upright, as without the root the tree will fall. Similarly, the tree data structure has a root that is the initial node of the tree from which the rest of the tree is built upon; the root node is identified in Figure 3-3.

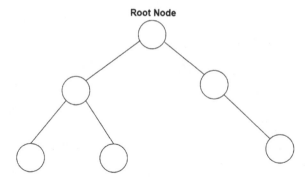

Figure 3-3. Root node

When a node is connected to another node in the direction away from the root, we call that node a *child* node. When the node is connected to another node in the direction toward the root, we call such a node the *parent* node. A parent can have multiple child nodes. Figure 3-4 shows a depiction of the parent and child nodes. However, a child node cannot have multiple parents. If a child node has multiple parents, then it is what we call a *graph*, a concept that we will examine in a later chapter.

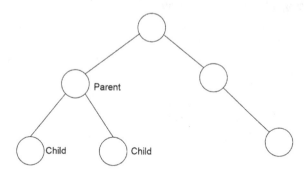

Figure 3-4. Parent and children

When we think about the structure of a tree, we know that the roots are connected to the trunk, the trunk is connected to the branches, the branches are connected to the leaves, and the leaves are connected to nothing. Likewise, in the tree data structure, we call the last nodes of a tree *leaves*, as shown in Figure 3-5. These leaves do not have any child nodes and are the last nodes on the tree.

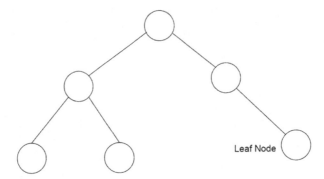

Figure 3-5. Leaf nodes

We call the links that connect the nodes within a tree the *edges* of the tree. A *subtree* refers to a tree that is the child of a node. Figure 3-6 shows these concepts. Here we distinguish the subtree by surrounding it with dotted lines.

Some other things you should know about trees is that nodes contain data with a key used to identify the data stored by the node and a value that is the data contained in the node. This relationship between a key and a value in the node is called a *key-value* type structure. You should also know that the process of navigating a tree is called *traversal*.

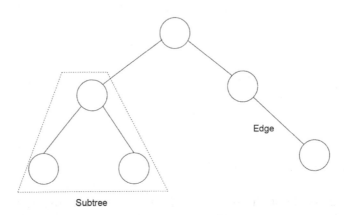

Figure 3-6. Subtree and edge

Binary Trees

The most used data structure is by far the binary tree. The binary tree is so named because each parent node can be linked to only two child nodes within the tree. The tree we have been looking at for the duration of the chapter is in fact a binary search tree. Figure 3-7 shows this tree again.

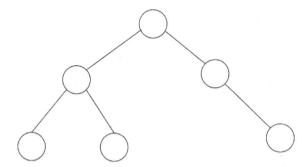

Figure 3-7. Example binary tree

The most common type of binary tree we will encounter is the binary search tree. Remember, we said previously that the nodes in a tree have a key-value structure. Well, binary search trees keep their keys sorted. In the binary search tree, all nodes are greater than the nodes in their left subtree; however, they are smaller than the nodes in their right subtree. Figure 3-8 shows this concept.

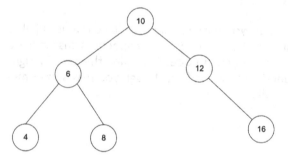

Figure 3-8. Binary search tree

The smallest node in a binary tree is located at the end of the leftmost sub-tree line stemming from the topmost node, whereas the largest node is located at the end of the rightmost subtree line stemming from the topmost node. In Figure 3-9, labels are placed on these nodes so that this is clear to you.

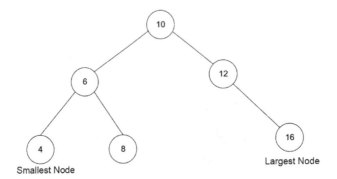

Figure 3-9. Largest and smallest nodes

There are three main things we do with binary search trees. We add nodes to the tree, we delete nodes from the tree, and we check nodes to see whether a key we are trying to locate is present. Binary search trees usage is extensive as they are efficient for storing sorted data based on their structure.

AVL Tree

For whatever reason, you may have a binary tree that is unbalanced. When we have an unbalanced tree, typically what happens is that many of the nodes in the tree have only one child node. Consider the tree in Figure 3-10; this is what an unbalanced tree looks like, though you may sometimes see it making a much longer chain.

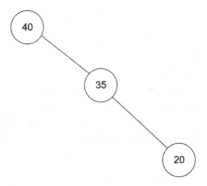

Figure 3-10. Unbalanced binary tree

To counteract this problem, we must perform what is called *balancing* on the tree. When we balance a tree, what we do is make the tree have the minimum height that is possible, while maintaining the characteristics we expect of the tree. Balancing a tree is important because of how the tree operates in memory; a balanced tree will allow for more efficient memory usage.

If we have such a tree, many data structures will become more efficient. Some binary search trees can balance themselves. For example, if we balance the tree, it should look like Figure 3-11.

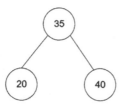

Figure 3-11. Balanced binary tree

One such tree is called the Adelson-Velsky and Landis (AVL) binary tree. Once the tree detects a height difference between two child subtrees, it performs what is known as a *tree rotation*. A tree rotation moves one node of the tree up and the other down. The method by which the tree performs balancing is beyond the scope of this book. However, what you should know about balancing trees and the process whereby a tree is balanced takes time, and as such there is a time complexity associated with each self-balancing tree. The AVL tree has a time complexity of O(log N). AVL trees, however, are used in some database retrieval systems.

There are other tree data structures, and we will look at the common ones for the remainder of this chapter.

Red-Black Tree

One simple type of binary search tree that is hard to forget is called the *red-black* binary search tree. The red-black binary search tree is like the AVL binary search tree in that it is self-balancing. However, because of its structure, it is more efficient than the AVL tree as it performs fewer rotations to balance the tree. The red-black tree also has a time complexity of O(log N).

There is an idiosyncrasy in that the nodes of the tree have a bit that is interpreted as red or black. The tree has a root node that is usually black, and each red node has children that are black. Figure 3-12 shows an example of a red-black binary tree.

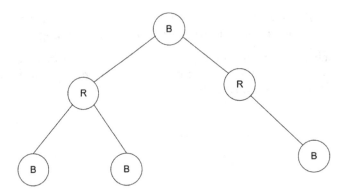

Figure 3-12. Red-black tree

B-trees

The B-tree is another data structure you may hear computer scientists talking about as it is used when designing database systems. A B-tree is a type of tree that has features for self-balancing. However, unlike the binary search tree, the B-tree has parents that can have more than two child nodes. For example, in Figure 3-13 we see a tree structure where two of the parent nodes have three child nodes. This would qualify this data structure as a B-tree.

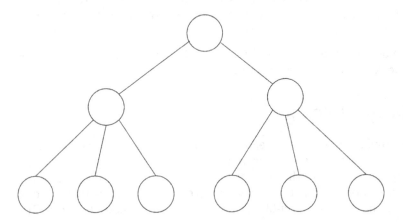

Figure 3-13. A B-tree

Many file systems utilize a B-tree for the data hierarchy. Think about it, if we have a file system, there will be a scenario where we have a folder in a file system, and the key-value structure of the nodes will allow us to associate each name of the folder to an object in the file system. Each folder may contain another folder, which may contain multiple files and so forth. As such, the B-tree is an ideal candidate to be applied in this scenario.

Heaps

Heaps are a type of tree-based data structure that you will need to be familiar with since you will come across them so often in real-world development. Heaps are useful in applications where we need quick access to the maximum and minimum nodes within a tree. Heaps are a type of binary tree data structure.

To do this, the heap implements a priority queue, which we discussed earlier. The heap can be structured in two ways. When the heap is structured in such a way that the root node is the highest value in the heap and each node is less than or equal to the value of its parent, we call it a *max heap*, as in Figure 3-14.

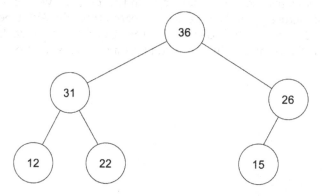

Figure 3-14. Max heap

However, if the root node has the minimum value in the heap and each node within the heap has a value greater than or equal to its parent, as in Figure 3-15, we call it a *min heap*.

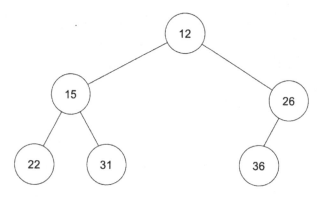

Figure 3-15. Min heap

Both min heap and max heap have their applications, and neither is superior to the other; it is up to the application designer or programmer to decide which is best for the algorithm they are designing.

There is something that you should know, which is that the heap data structure is not the same as *heap memory*. In fact, the heap memory is implemented in a much different way than the heap data structure. This is something that many programmers who are learning about heaps confuse, so it is important to make this distinction.

Conclusion

This chapter was packed with knowledge. In this chapter, we looked at tree-based data structures including the binary search tree, the AVL tree, the red-black tree, and B-trees. We also looked at the heap data structure. In the next chapter, we will look at another important data structure, that of hash-based structures.

Hash Data Structures

In the previous chapter, we looked at trees. Trees are important to understanding some complex topics in computer science, and we will use them going forward. In this chapter, we will look at a data structure that is equally important to the tree. This type of data structure is the *hash table*. Hash tables utilize hash functions to operate. In computer security in particular, cryptography and checksums rely heavily on hashing. In this chapter, we will look at the basics of hashes, hash functions, and hash tables.

Hashes and Hash Functions

Before we delve into hash data structures, we must look at what hashes are. When we are talking about a hash, what we mean is that there is some black box we call a *function* that converts one value to another. In Chapter 1 we talked about functions, so you may refer to that chapter for a further description.

While this may seem basic, that's all a hash is—a function that converts one value to another. To understand a hash function, it may be easier to look at how it works in a step-by-step manner. We will start our discussion by looking at the function.

© Armstrong Subero 2020
A. Subero, *Codeless Data Structures and Algorithms,*
https://doi.org/10.1007/978-1-4842-5725-8_4

Let's imagine we have a function in the form of a box; let's call it a *function box*. Now imagine we take our function box, and we input the letter A. Don't worry about how the box functions; just follow along with the logic.

Now within our box we have some special gizmos; every time we input the letter A, we get the number 1 as an output. We can see what such a setup looks like in Figure 4-1. For all its complexity, this is what the hash function does; we input data and get an output called a *hash value*.

Figure 4-1. Our function box

We'll take this a step further. In Figure 4-2, we replace our function box with a hash function. The function is like before; we input the letter A and get an output. However, when we input data into the hash function instead of getting the number 1 as an output, the output is a hexadecimal sequence. This is how the most commonly encountered hash functions work.

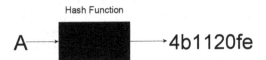

Figure 4-2. A hash function

The hash function is special in that it does not matter if the data being input is a single character, a string of characters, or even symbols. The hash value will always have the same fixed size.

Hash functions differ from regular functions. For a regular mathematical function, an element in the domain is mapped to exactly one element in the range. However, in a hash function, there is a small probability that two differing inputs may produce the same hash value. When this occurs, we get a hash collision. However, this is rare because even a change of one of the input bits to the hash function can result in a drastically different output of the hash function.

Good hash functions will have a value that is easy to compute and will avoid collisions. In the real world, there are many hash functions you will encounter, and many of them will have been designed for you.

Hash Table

Now that we understand how hashes function, we can understand the hash table data structure. Hash tables, like their name implies, utilize hashing, and the hash function is a fundamental part of the hash table data structure.

Hash tables, for all their complexity, can be simplified into this one-line description:

> At its core a hash table is a key-value lookup system.

That's it. That's all a hash table is, and we will keep this in mind going forward.

When we say that the hash table is a *key-value lookup system*, this means that in the hash table, every key has a value associated with it. This arrangement of data drastically increases the speed we perform lookups.

Since each key has a value associated with it, once we know the key, we can find the value instantly. This means that hash tables will have a time complexity when searching for an element of $O(1)$.

This is advantageous because when we want to retrieve a value from, let's say a database table, it can happen in most cases instantly, which the program end user will greatly appreciate.

The hash table utilizes hash functions to perform the lookup. The key, which is usually a string, has a hash function applied to it; this then generates the hash value that is mapped to an index in the data structure (an array in a basic hash table) that's used for storage. This means we can use the generated hash value to calculate the index in the array that we want to find or add to the table.

This might seem a little confusing, so an illustration of what I am saying will help; see Figure 4-3. Let's imagine we have three strings, from STR0 to STR2. The hash function will take each string and have it mapped to an index in an array thanks to our hash function.

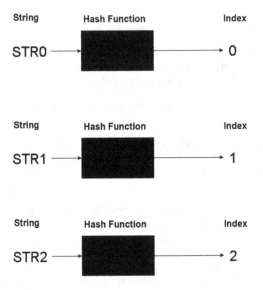

Figure 4-3. Hash function

In the end, after each string has a hash function applied to it and generates an index for each string, our hash table will look like Figure 4-4.

INDEX	STRING
0	STR0
1	STR1
2	STR2

Figure 4-4. Hash table

While this method is efficient, there is a major drawback. Using an array, due to the underlying implementation of hash values and array sizes (that, honestly, we don't need to understand how to use hash tables), can lead to a hash collision.

To solve this problem of a hash collision, the hash table implements what is known *chaining* to prevent this from happening.

Chaining is an innovative idea that rather than storing elements in a simple array of elements, the hash table will store it as an array of elements that are linked lists.

Confused? Well, maybe some illustrations will clear it up. In a regular array-based linked list, each key element has a value associated with it. It would look something like Figure 4-5.

Index	String
0	STR0
I	STRI
2	STR2

Figure 4-5. Regular array linked list structure

The chaining method will have each element in the array as a liked list, as in Figure 4-6.

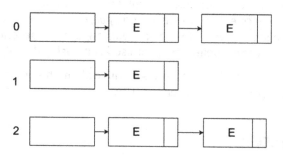

Figure 4-6. Chaining method

As shown in the figure, instead of each element in the table simply having a value assigned to it, each element is a linked list that has items labeled as *E*, as in the figure. While this may not seem to have any added benefit, it helps with collision resolution.

The chain method resolves collisions by a simple method. If the hash function returns the same hash value for multiple elements, we simply "chain" those elements together with the same hash values at the same index in the hash table.

At index 0 and index 2, you see multiple items in a list associated with an element, which is because of the collision resolution ability of the hash table.

While this solves the problem of a hash collision, as the chain length increases, the time complexity of the hash table lookup can worsen.

Computer Security Basics

Most commonly you will encounter hashing in computer security applications. However, before you understand exactly where hashing fits within this realm, it is important to have a basic understanding of computer security, which I will briefly cover in this section.

Computer networking is an essential part of computing as it allows computers to share resources. Exchanging data and information has allowed us to have a globalized society. Through the Internet, e-mail, and the Web people can work, study, and even form relationships with people all over the globe.

However, when sharing information, sometimes we do not want the information to fall into undesirable hands. Take, for example, Internet banking. When we do Internet banking, for example, we do not want our account information and passwords to be intercepted by a third party.

When sharing information, sometimes a third party can pretend to be someone they are not. Within computer security, when someone tries to identify themselves as someone else, this is called *spoofing*. To prevent spoofing and other compromises of the information exchange, we use encryption. The major way we avoid data being compromised is by using digital signatures.

Before we get into digital signatures, we will look at the cryptosystems that form an essential part of computer security.

Cryptosystems

Within the context of computer security, a cryptosystem is a set of algorithms that convert the input to the encryption process called *plaintext* into the output of the cryptosystem called *ciphertext*. When you convert plaintext into ciphertext, this is called *encryption*, and the conversion of ciphertext into plaintext is called *decryption*.

Cryptosystems use something called a *key* to aid the algorithm for the encryption process. A shared key cryptosystem uses the same key for encryption and decryption. How it works is that the person who wants to send data to someone else encrypts the data using a key. This key will convert the data into ciphertext. The receiver uses the key to decrypt the ciphertext and reobtain the original data.

While the shared key method seems secure, there are problems with this method. If a third party intercepts the encrypted message, since they do not have the key, they will not be able to decrypt the message. A problem arises if the key is sent over a network from one party to another, then the third party can intercept it, obtain the key, and decrypt the message.

This key distribution problem highlights the major flaws of the shared key cryptosystem. To aid with this problem, the public key cryptosystem was developed.

Public-Key Cryptosystem

In the public-key cryptosystem, different keys are used for encryption and decryption. We call the key that is used for encryption a *public key* and the key that is used for decryption a *secret key*. If the sender transmits the public key along with the data, there is no problem as the public key can only encrypt data and not decrypt it.

Hashing vs. Encryption

With all this talk about computer security and cryptosystems, you may be wondering how hashing fits into all this. It is important to make the distinction between hashing and encryption.

Encryption is the name given to the process of making information unreadable by muddling it so that only someone with a key can use it. Within the encryption process we use an encryption algorithm called a *cipher*. Encryption assumes that we will use the cipher to encrypt and decrypt information, which is to say we expect to recover the information that we originally encrypted.

Hashing, on the other hand, takes data as an input and produces a fixed-length output without the intention of obtaining the original data. Whereas encryption is a two-way process, hashing is a one-way process.

As simple as it seems, it is important to keep this in mind and not to confuse hashing with encryption.

Role of Hashes in Computer Security

Within computer security, hashes have several uses including digital signatures and user authentication, which we will discuss in this section.

The first prominent role you will encounter for hashes is in digital signatures. Digital signatures are used to validate the legitimacy of digital data. In other words, it is a way of verifying the data received came from who it was supposed to come from. To do this, the data must be signed by the holder of the private key.

Two commonly encountered digital signature methods are Rivest-Shamir-Adleman (RSA) and Digital Signature Algorithm (DSA). These digital signature methods are similar with the exception that DSA can be used only for creation and verification of signatures but cannot be used for encryption. RSA, on the other hand, can be used for both digital signatures and encryption.

Within both methods, hashes are employed to aid with ensuring the signatures are secure. Within the RSA signature method, the process of signing involves using a hash function on the data before it is encrypted. The DSA method uses a cryptographic hash function based on the Secure Hash Algorithm (SHA) to aid in the process.

Hashes also find uses in user authentication methods such as in password applications. Since hashes are one-way and 1:1 functions, this makes them ideal for using with password security.

When we input a password into a login form, for example, it is verified with a password that is stored in a database that is linked to the application. If security is breached and someone gains access to the database, if the password is stored as plain text, then a malicious hacker can see all the passwords.

Now, if the password is hashed, then even if security is compromised, the passwords remain secure. From the time the user creates the password, we will be able to store it securely without knowing what it is. This is because the output of the hash will be stored and not the plain text. This hash is consistent, so it can be stored and compared without knowing exactly what it is.

There are other uses of hashes within the computer security realm including random number generation, message authentication code (MAC), one-way functions, and any other application the cryptographic engineer may fathom.

Hashes and Cyclic Redundancy Check

With the advent of the Internet of Things, you might encounter some embedded system that may have a microcontroller within it. One of the modules you may encounter on a microcontroller system is one used for the cyclic redundancy check (CRC), which uses hashing for operation.

A cyclic redundancy check is a method of detecting errors in digital data. To do this, a hash function is used to verify the authenticity of data. The CRC works by taking a checksum, which is a fixed number of bits and attaching it to the message being transmitted using the hash function. On the receiving end, the data is checked for errors by finding the remainder of the polynomial division of the contents transmitted.

If the data is checked and the CRC does not match the sent CRC, then the data is most likely corrupted. The application designer can then choose to have the data ignored or re-sent by the sender.

The CRC module finds extensive use in embedded systems for transmitting digital data via Ethernet and WiFi, for example.

Other Hash Uses

Some applications such as search engines, compilers, and databases must perform a lot of complex lookup operations. These lookup operations are time intensive and time critical, and regular data structures like linked lists may not be able to keep up.

As such, we need some operation that will perform in constant time, and as we stated earlier, hash tables typically have a time complexity of $O(1)$. This fast constant time is ideal for such applications. This is where hashes shine and are used in such applications.

Conclusion

In this chapter, we examined the basics of hashing, hash functions, and hash tables. In the next chapter, we will look at graphs, which are an important concept in artificial intelligence that utilize graphs and graph searching algorithms for problem-solving. Should you need more information on the topics in this chapter, especially if your interest in computer security has been awakened, the book I personally recommend is *Practical Information Security Management* by Tony Campbell (Apress, 2016). You can also check several web sources on the topic.

Graphs

In the previous chapter, we looked at hashes, hashing, and hash tables. In this chapter, I will introduce an important mathematical concept that we will need to understand to comprehend the operation of later algorithms. That concept is graph theory. Graph theory can become involved and complex; however, for the sake of understanding, we will take the 20,000-foot view and avoid getting bogged down in the details. Graphs are essential to many parts of computer science, and once you understand how they work, you will be able to apply them to solve a wide array of problems. Mathematical gurus need not read this chapter, but for anyone else, it's worthwhile to understand the basic concepts I present here.

Dimensions, Points, and Lines

Before we delve into graph theory, I think we should review three important mathematical concepts, that of the dimension, the point, and the line. Yes, I know that you already learned about them in high school; however, they form a solid base for us to build upon. While these seem elementary and unrelated to the topic at hand, they are essential to understanding graphs in computer science.

When discussing the point and the line, I think it's best we do so from the perspective of a mathematical dimension. In mathematics, when we refer to a dimension, what we are referring to is the distance in a direction. This is something you should be vaguely familiar with as when we are talking about distance, this distance may be length, width, or height.

© Armstrong Subero 2020
A. Subero, *Codeless Data Structures and Algorithms*,
https://doi.org/10.1007/978-1-4842-5725-8_5

Common geometrical shapes have several dimensions with the most common having two or three dimensions. Two-dimensional shapes have the dimensions of length and height, and three-dimensional shapes have all three dimensions of length, width, and height. With this in mind, let's move forward.

A point is a unique concept, because a point has no dimensions, only a position. Remember that we said a dimension is a distance in a direction. Well, you can think of a point as the starting position. To know where points are, we use a dot to represent them. Figure 5-1 shows a point.

Figure 5-1. A point

While a point exists with no dimensions, at the first dimension we have a line. A line is a point that has simply been moved in one direction. Figure 5-2 shows a line.

Figure 5-2. A line

When we have two lines meeting each other, we get what is called a *vertex*; the plural form of vertex is vertices. In Figure 5-3, the point where the two lines meet is the vertex.

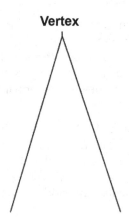

Vertex

Figure 5-3. A vertex

While this knowledge comes from basic geometry, it has very real applications in DSA as it provides the foundation for us to understand more complex topics, as we will see in the next section.

Graphs

Computers are all about connections; in fact, computer networking is all about connecting computers. From this basic concept of connecting things comes the graph. What a graph does is show the connections between nodes. Graphs are important because they provide a visual means for us to see the connections between objects.

Graphs are a kind of overlapping gray area where you can really see the connection between computer science and mathematics. This branch of mathematics is called *graph theory*, and it has a lot of applications in computer science.

Graphs vs. Trees

The best way to think about graphs is to compare them to something we already know about. In this case, since we already examined trees in an earlier chapter, we can think of the graph as a tree with no root; additionally, a graph has no nodes that can be identified as a parent or child. We can also think of it as a chaotic graph where each child node has multiple parents. Figure 5-4 shows a tree data structure.

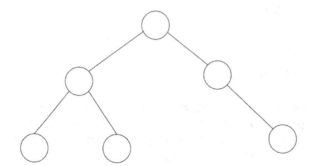

Figure 5-4. Tree data structure

While the tree feels very structured, the graph feels more expressive and closer to a real-world representation of connections because unlike trees, which can represent multiple paths between only two nodes, graphs can give a better model of the representation of real world connections. In real life, living and nonliving objects have multiple interactions with many things, and graphs can more accurately represent and model such relationships.

Some people like to think about trees as a kind of minimalist graph. This is because in real-world development, most of the algorithms that will work on graphs will work on trees because trees are basically graphs without any cycles or cyclic interaction taking place.

More About Graphs

The nodes on a graph are sometimes called *objects* or *vertices*. The links connecting the vertices are known as *edges*. Vertices that are connected by an edge are said to be adjacent to another one.

These edges may be directed from one node to another; when this happens, we call this graph a *directed graph*. It is easy to identify directed graphs with the arrows on their edges.

An undirected graph is one in which there are no directed edges, and because there are no directed edges, you can traverse edges both ways between the nodes. Figure 5-5 shows an undirected graph.

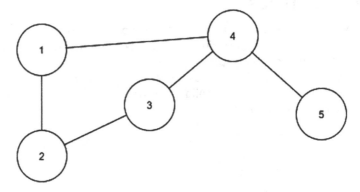

Figure 5-5. An undirected graph

A directed graph is one in which the edges are directed, which is to say the edges connecting the vertices on the graph each have a direction associated with them. We call these edges *directed edges*, and sometimes we call the directed edges *arrows*. These directed graphs are also called *digraphs*. Figure 5-6 shows us a directed graph.

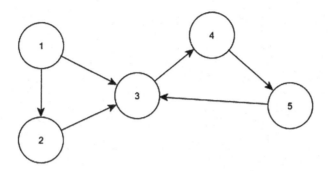

Figure 5-6. A directed graph

To get from one of the vertices of the graph to another, we follow what is known as a *path* along the edges of the graph. If we take a vertex of a graph and draw an edge from that vertex to itself, we get what is called a *loop*. In a loop, the initial and final vertices coincide.

When a graph exists such that it is a graph that exists within a larger graph, we call that graph a *subgraph*.

Basically, that's all a graph is, a bunch of circles being connected by a bunch of lines. What makes graphs useful is that they allow us to see the relationship between objects (nodes). This property of graphs makes them useful for many things, and as we will see later, a lot of algorithms based on searching are dependent on graphs.

Weighted Graphs

We have already established that graph theory is all about connecting a bunch of circles to a bunch of lines. The edges, as we know, are the links connecting the nodes on the graph.

Sometimes these edges have a weight. This means there is a number of values associated with each edge on the graph. When edges have a weight associated with them, we call this type of graph a *weighted graph*. Figure 5-7 shows us what a directed graph looks like.

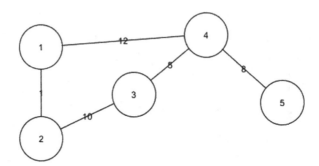

Figure 5-7. A weighted graph

While the graph in Figure 5-7 is an undirected graph, it is important to note that weighted graphs may also be directed. Weighted graphs find a lot of applications in computer science, and many algorithms are dependent on weighted graphs.

Graphs and Social Networking Applications

While the theory presented in this chapter may seem dull and boring, there is a way to see the importance of graphs within real-world applications by looking at how graphs will operate within a program designed as a social networking site.

On a social networking site, someone will have a profile. This profile will contain personal information about someone, such as where they grew up, their age, pictures of them, and where they went to school.

We can think of this profile of a person as a node on a graph. Let's look at an example person named John; John is a lone node, as shown in Figure 5-8.

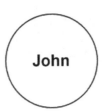

Figure 5-8. John node

If we had only John on our site, then it wouldn't be a networking site, would it? It would be a web page about John. To keep the site from being all about John, each person on a social networking site will be friends with another person on the site. Let's say John and Alex both went to the same elementary school and connected online and are now friends with each other on the application. They would them both be connected nodes on the graph, as in Figure 5-9.

Figure 5-9. John and Alex connection

Alex is friends with Sue, and Sue now joins the graph, as in Figure 5-10, as the circle of friends keeps growing.

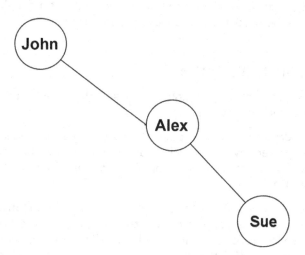

Figure 5-10. Growing circle of friends

Now imagine Alex meets Jill at a business meeting, Jill and John meet at a dinner party, and all of them are now friends with each other on the application. Our graph of connections will now resemble something like in Figure 5-11.

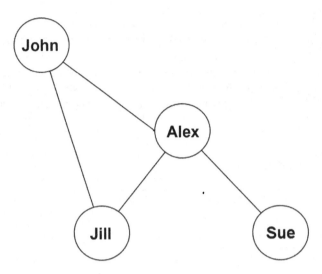

Figure 5-11. The circle keeps growing.

In real-world social networking applications, variants of a simple graph just like this one can be used as the basis of the application.

The Graph Database

Another direct application of graphs and graph theory is in the design and mechanics of the graph database.

When storing data, we have a choice between using a file system or using a database management system. Storing data in a file system has its drawbacks as it is limited in operation, lacks uniformity in terms of format, leads to data duplication, and lacks security.

To fix the problems of a file system, the relational database management system (RDBMS) was created, which fixes the problems of the file system by allowing data to be stored in tables consisting of rows and columns. There is also a schema in the DBMS that describes the data being stored as well as the attributes of the table within the database.

What makes the relational database system useful is that there is a key system in place that allows tables to have relationships with one another as each table has a primary key, and relationships between tables are set using foreign keys.

A problem with relational databases arises when scaling the database. The computing time of some operations between connected tables can be very computing intensive. Graph databases fix many of the problems associated with the RDBMS by using a graph data structure for organization. There are many graph databases for different functions. Should you need more information on graph databases, I recommend the book *Practical Neo4j* by Gregory Jordan (Apress, 2014).

Conclusion

In this chapter, we took a brief look at graphs. We reviewed the basics of dimensions, points, and lines. We also discussed graphs and touched on weighted graphs. We briefly talked about two applications of graphs in social networking applications and databases. In the next chapter, we will delve into algorithms as we cover linear search and binary search.

Algorithms

Linear and Binary Search

In the previous chapters, we looked at data structures. At this point, you have more knowledge of data structures than a lot of people I have spoken to. While you may not know them inside out, you still have enough knowledge to at least know the purpose of each major data structure and may be able to apply them to solve your problems. In this chapter, we will deviate from looking solely at data structures and look at algorithms as well.

Mathematical Concepts

Prior to beginning our discussion on algorithms, there are a few mathematical concepts we must look at. These concepts will form the basis of understanding algorithms later when we discuss them.

© Armstrong Subero 2020
A. Subero, *Codeless Data Structures and Algorithms*,
https://doi.org/10.1007/978-1-4842-5725-8_6

Linearity

Before we get to any other discussions concerning algorithms, we can take a little time and talk about the concept of linearity. Figure 6-1 shows a line.

Figure 6-1. A line

In mathematics, when we say something is *linear*, we mean it can be represented on a graph as a straight line. Figure 6-2 shows a linear graph.

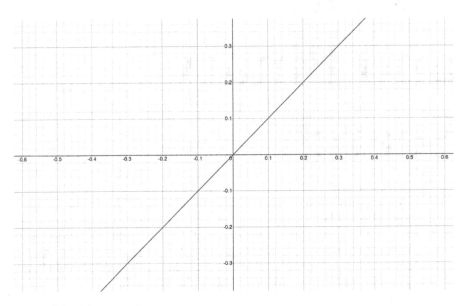

Figure 6-2. A linear graph

Logarithms

Logarithms are a simple concept that people have made overly complex. We will begin our discussion of logarithms by discussing exponents.

An *exponent* is a way to represent a mathematical operation. An exponent consists of two numbers; one number is called the *base*, and the other is called the *exponent* or *power*, as in Figure 6-3.

$$e^x$$

Figure 6-3. An exponent

In Figure 6-3 we see an example of an exponent; the base is *e*, and *x* is the power or exponent.

The letter *e* has a special meaning in the realm of mathematics and by extension computing. *e* is known as Euler's number, and it the base of natural logarithms. This important number is used in many applications from half-life calculations to economics and finance.

Figure 6-4 shows an exponential graph. This graph has a common plot and is easily recognizable.

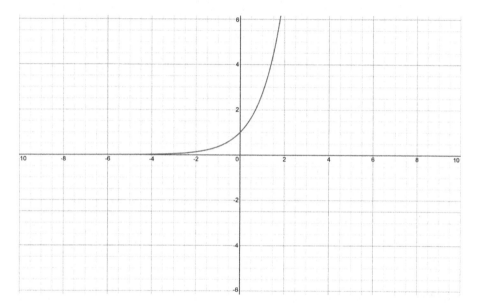

Figure 6-4. An exponential graph

At the core of many mathematical concepts lies a function or operation and its inverse. Consider the four basic mathematical operations of addition, subtraction, multiplication, and division. Addition is an operation whose inverse is subtraction, and similarly, multiplication is an operation whose inverse is division.

Since each of the basic operations has an inverse, it's only fair that other complex functions also have inverses. For exponential functions, the inverse is the logarithm.

Logarithms are the reverse of exponents. To understand how this inverse works, we can look at the statement in Figure 6-5.

$$If\ y = b^x\ then\ x = log_b\text{y}$$

Figure 6-5. Logs and exponents relationship

Logarithms find a lot of uses in computer science and engineering. As you will come to understand, many algorithms perform in a logarithmic manner. Figure 6-6 shows the logarithmic graph.

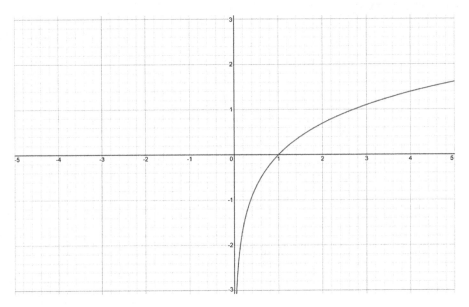

Figure 6-6. A logarithmic graph

The logarithmic graph also has a distinctive shape, and this is something that you should learn and be able to recognize as you will encounter it throughout your study of DSA. At this point, we have a little understanding of linearity, exponents, and logarithms. With this knowledge, we begin our discussion of the searching algorithms.

Linear Search

Searching for something seems like a trivial task for us. Let's say that we have an array of elements from 1 to 10.

[1, 2, 3, 4, 5, 6, 7, 8, 9, 10]

Now let's say we wanted to find the number 7. How would we do that? Well, for us humans, it is easy to look at the array and simply see the number 7 and identify that it's at index 6. For a computer, however, we must devise some algorithm to find the number 7. We could do something like this:

1. Have a variable that holds the number 7.

2. Loop through the array.

3. Compare each element in the array to our variable that stored the number 7.

4. If the variable matches an element in our array, we have found our number.

5. Return the index of the number we found.

This simple algorithm seems like a good way to find the number we need. In just five steps, we can find the index of the number we need. However, let's say that we are looking for the number 10,000 in an array of numbers from 0 to 10,000. The time would increase significantly. In fact, if we keep adding elements to our array, the runtime will increase, and when we reach 10,000,000,000 numbers and we are looking for the number 9,999,999,999, the algorithm will seem like it isn't running at all!

We call this type of algorithm a *linear algorithm* having the notation $O(n)$ since as the number of input elements increases, the runtime of the algorithm increases directly proportional to the increase in the number of elements. Think back to our talk on linearity, and you will realize that this is in fact a linear way of searching for an element.

The linear search is slow because it must scan all elements in the input to find the answer. For small array sizes, this may not be an issue; however, as the size increases, the algorithm slows down to the point where it will take hours or even days to find the number we are looking for.

However, there may be a situation where the linear search algorithm seems fast and may fool the uninformed programmer.

Let's say our list of numbers is ordered as follows:

[7, 5, 3, 10, 6, 2, 1, 8, 4, 9]

In this scenario, if we try to find the number 7 as we previously did, using the same algorithm, it will return a hit on the first try! The number 7 will be returned instantly! In fact, if we wrote our previous linear search algorithm and it gave such a result, we might be inclined to believe that the algorithm is instantaneous, having a time of $O(1)$.

Don't be fooled by this. The algorithm still has a time complexity of O(n). To prove this, let's take the same array and swap the first and last elements.

[9, 5, 3, 10, 6, 2, 1, 8, 4, 7]

Will the hit still be instantaneous? No, it will not, as the search algorithm will go through each element one by one until we find the one that we need. The time complexity will still be O(n) as the algorithm is still linear.

The linear searching not only can be applied to numbers but also to names, images, audio clips, and any type of data we may be working with. Searching is such an important thing that many algorithms have been devised to make searching as fast as possible for differing data structures.

While the linear search seems simple, it is inefficient. We will look at another searching technique that employs a logarithmic searching technique, the binary search.

Binary Search

While linear search is simple to implement, if suffers from one flaw. As the number of input elements to the searching algorithm increases, so too does the runtime of the algorithm increase. To combat this problem, a brilliant algorithm called *binary search* solves this problem.

We will learn how binary search works by walking through its method step-by-step.

Let's say we have an array with numbers from 1 to 9.

$$[1, 2, 3, 4, 5, 6, 7, 8, 9]$$

If we were looking for the number 6, we could begin by identifying the center number of the array, which is 5.

[1, 2, 3, 4, **5**, 6, 7, 8, 9]

We will use the elimination method and eliminate numbers that are lower than our target number. To do this, we compare 5 to 6. Since 5 is less than 6, we know that the numbers less than 5, inclusive of 5, will not contain our target number. So, we eliminate 1, 2, 3, 4, and 5 until we are left with the following array:

[6, 7, 8, 9]

Now that we have this array, we will repeat the process we previously did. We will try to locate the number 6. We find the center number of the array, which we identify as 7.

> [6, **7**, 8, 9]

We compare the 7 with the number 6. Since 7 is greater than our target number, we know that any number larger than 7, inclusive of 7, would be too large. So, we can eliminate these numbers. We are left with this:

> [6]

After we perform the second elimination, we are left with our number that we wanted to find. We can then identify the index of the number we are looking for (since we know where it is) and return that index.

Though binary search seems like a longer and more involved method, it is actually a simple algorithm that is powerful.

We use the simple concept of taking an array, and we keep dividing that array in half and do a comparison to determine whether we have our target number.

The binary search is much more efficient than the linear search having a time complexity of $O(\log n)$. In fact, the algorithm is second in speed only to the constant time complexity $O(1)$. Because as we know with logarithmic time complexity, as time increases linearly, then n goes up exponentially. This means that more elements will take less time to search.

When we are performing a simple search, the speed gains seem trivial. However, as we start performing more complex searching, the gains will be significant.

The only caveat with using binary search is that it works only if the array is sorted. If the array is not sorted, then binary search will not work.

With the power of binary search, you can perform a lot of searching tasks quite easily and in just a few steps. This algorithm is nice to use because it is both simple and powerful.

Conclusion

In this chapter, we got an introduction to algorithms, and we looked at two fundamental algorithms for searching, the linear search algorithm and the binary search algorithm. In the process of learning about these algorithms, we discussed the mathematical concepts of linearity and logarithms. In the subsequent chapters, we will discuss more algorithms that can be practically applied to your programs.

Sorting Algorithms

In the previous chapters, we looked at searching, and we explored linear search and binary search. In this chapter, we will look at another set of important algorithms—sorting algorithms. Sorting is a common task for many algorithms as many applications may need to search through text or images in various forms. We will cover all the commonly encountered sorting algorithms in this chapter.

Sorting

The sorting algorithms are important. This is because many computing algorithms require that the items they are working on have some type of sorting in place. When we talk about sorting, what we are really talking about is arranging the data in a particular way.

We can arrange the data either by category, which is to say the items have properties that are similar, or with an order or sequence.

We already saw an example of binary search, which requires the array be sorted before the information is input. There are other situations that may require that the data be sorted. Think of when you are running a query on a database and you want to order items by certain properties.

© Armstrong Subero 2020
A. Subero, *Codeless Data Structures and Algorithms*,
https://doi.org/10.1007/978-1-4842-5725-8_7

Sorting can also be used to identify duplicate data because once the data is sorted, the algorithm will quickly identify the duplicate data. There is also the case that if the data is sorted, it will be very fast to locate information.

In the following sections, we will look at common sorting algorithms I have seen used in programs. We won't go into the integral details of each; however, we will cover just enough of each algorithm that you will have a simplistic understanding of its operation. The reason I won't go into the nitty-gritty details is because many languages have sorting algorithms provided as library functions. However, having some knowledge of how these libraries work under the hood can help make your program design better.

Bubble Sort

The first sort we will look at is the simplest sorting algorithm to understand, bubble sort. We will walk through the bubble sort step-by-step to show you how it works. Imagine we have an array of even numbers as follows:

[10, 18, 6, 2, 4, 16, 8, 14, 12]

The bubble sort algorithm will look at the two numbers to the far right of the array, which are 14 and 12.

[10, 18, 6, 2, 4, 16, 8, **14, 12**]

The sorting algorithm will compare these two numbers and will determine that the smaller number on the right, 12, is smaller than 14. The numbers will be swapped, and the new array will be as follows:

[10, 18, 6, 2, 4, 16, 8, **12, 14**]

The algorithm will then move one place to the left and compare the next two numbers, which are 8 and 12.

[10, 18, 6, 2, 4, 16, **8, 12**, 14]

After the comparison, the algorithm will determine that 12 is greater than the 8; therefore, the algorithm won't perform any swap.

[10, 18, 6, 2, 4, 16, 8, 12, 14]

The scale will keep moving one position to the left and comparing numbers until the end of the sequence is reached. After doing this operation, the array will look something like this:

[2, 10, 18, 6, 4, 8, 16, 12, 14]

After the algorithms are processed, the smallest value is moved to the left-most edge of the array, and the number on the end, 2, is considered sorted, and the algorithm returns to the far right of the array.

[2, 10, 18, 6, 4, 8, 16, **12, 14**]

The algorithm keeps going over the array until the array is sorted. After the algorithm is complete, we have our sorted array.

[2, 4, 6, 8, 10, 12, 14, 16, 18]

The bubble sort is an inefficient algorithm with a complexity of O(n^2). Despite its simplicity, the slowness makes it somewhat unfeasible for any real-world applications; as such, there are better searching algorithms, which we will discuss in the forthcoming sections.

Selection Sort

In the previous chapter, we discussed linear search. One application of linear search is in the selection sort. Selection sort works as follows. Let's say we have an array of even numbers.

[12, 4, 14, 16, 18, 6, 10, 8, 2]

The selection sort will first use a linear search to locate the smallest value in the array. In this case, it's 2.

[12, 4, 14, 16, 18, 6, 10, 8, **2**]

The selection sort will take this small value and perform a swap between that number and the number to the furthest left of the array, in this case 12.

[**2**, 4, 14, 16, 18, 6, 10, 8, **12**]

The selection sort then keeps performing this operation until all the numbers are sorted.

[2, 4, 6, 8, 10, 12, 14, 16, 18]

While this algorithm is rather simple to understand, it has a time complexity of O(n^2) as it uses two loops to perform its function.

Since it uses the linear search to initially locate the smallest element, the algorithm runs more slowly as the number of input elements increases.

While selection sort gives slightly better performance than bubble sort and performs well on a small array of elements, it is still rather slow and inefficient for large arrays. For this reason, there are other sorting algorithms we will look at that even further improve this process.

Insertion Sort

Insertion sort is a clever sorting algorithm that is widely used. Insertion sort works in the following way.

If we take a list of unsorted even numbers:

[10, 6, 8, 14, 4, 8, 12, 18, 2]

the insertion sort will consider the number to the farthest left, 10, to need no further sorting.

[**10**, 6, 8, 14, 4, 8, 12, 18, 2]

The sorting algorithm will then look at the penultimate number farthest to the left and compare it to the number that is already selected as needing no further sorting. In this case, 10 will be compared to 6.

[**10**, **6**, 8, 14, 4, 16, 12, 18, 2]

Since 10 is larger than 6, the numbers are swapped. After the swap, the algorithm considers 6 to be sorted.

[**6**, **10**, 8, 14, 4, 16, 12, 18, 2]

The algorithm will then continue removing the number farthest to the left and comparing it to the numbers next to it. When the algorithm reaches 8, it will compare it to 10.

[6, **10**, **8**, 14, 4, 16, 12, 18, 2]

Since 10 is greater than 8, the numbers will be swapped.

[6, **8**, **10**, 14, 4, 16, 12, 18, 2]

After 10 and 8 are compared, the algorithm will compare 8 and 6. Since 8 is larger than 6, the algorithm won't move the 8 any further. This continues until all the numbers are sorted.

[2, 4, 6, 8, 10, 12, 14, 16, 18]

Since the insertion sort doesn't need to iterate over all the numbers in the array to find the smallest like selection sort, the algorithm, having a worst-case time complexity of O(n^2), is more efficient than selection sort.

Merge Sort

An ingenious method of performing a sort is with a merge sort operation. Merge sort is a type of sorting algorithm that uses the method of halving to perform the sort. This method of halving is called the *divide-and-conquer* method, and it is an important concept.

Let's imagine we have an array of unsorted even numbers.

[12, 8, 6, 14, 10, 2, 4]

The first process in the merge sort will be to divide this array in half.

[12, 8, 6, 14] [10, 2, 4]

After this first division into halves, we will divide these arrays into halves again.

[12, 8] [6, 14] [10, 2] [4]

We then divide it again until all the elements exist in an array by itself.

[12] [8] [6] [14] [10] [2] [4]

The algorithm will then take these individual arrays and combine them. They are combined in such a way that they are ordered from smallest to largest.

[8, 12] [6, 14] [2, 10] [4]

The algorithm will then perform a second combination stage where they are grouped into a larger combined array, ordered from smallest to largest.

[6, 8, 12, 14] [2, 4, 10]

Finally, the algorithm will combine these two arrays into one large combined array that has the elements fully sorted.

[2, 4, 6, 8, 10, 12, 14]

From this algorithm, we can see how merge sort operates; we divide our array into halves until they cannot be divided, and then we recombine them into halves until there is no other combination to take place.

Since in practice merge sort typically uses recursion to continually divide the input, the time complexity of the algorithm will be dependent on the number of times recursion has to take place, and typically this gives the algorithm a time complexity of O(n log n).

Quick Sort

Another sorting algorithm like merge sort that also uses the divide-and-conquer method for sorting is quick sort. Let's imagine we have an array of even numbers.

[6, 10, 16, 2, 4, 18, 8, 14, 12]

The quick sort algorithm works by choosing a number from the array called a *pivot*. This pivot number is usually chosen at random, though sometimes the algorithm designed may choose the characteristics of the pivot based on their preference. The algorithm will then take the rightmost number and leftmost number and place markers on them.

If our pivot number (P) is 4, then the left marker (LM) will be 6, and the right marker (RM) will be 12.

LM   P   RM

[**6**, 10, 16, 2, **4**, 18, 8, 14, **12**]

Once the algorithm has these numbers, the left marker will move to the right and select the first number that is greater than or the same as the pivot number. Similarly, the right marker will move to the left and select the first number it meets that is less than the pivot number.

LM   RM P

[**6**, 10, 16, **2**, **4**, 18, 8, 14, 12]

In this case, the left marker will select the 6, and the right marker will select the 2. Once the LM and RM select their numbers, they swap them.

LM   RM P

[**2**, 10, 16, **6**, **4**, 18, 8, 14, 12]

This act of using markers to find numbers and then swapping them allows numbers less than the pivot to gather to the left of the array and allows numbers that are greater than the pivot to gather to the right of the array.

Once this has occurred and the numbers are sorted in respect to being larger than or smaller than the pivot, the algorithm will treat them as halves. Once the halves are selected, the algorithm will do the same thing again, first with the numbers left of the pivot and then with the numbers right of the pivot. The algorithm will then keep sorting these smaller arrays until the array is completely sorted. The quick sort algorithm has a time complexity of (n log n).

Conclusion

In this chapter, we looked at the fundamental sorting algorithms. We looked at how to use bubble sort, selection sort, insertion sort, merge sort, and quick sort. In the next chapter, we will look at the complement of sorting, which is searching, and we will briefly discuss the searching algorithms.

Searching Algorithms

In the previous chapter, we looked at sorting. In this chapter, we will look at another equally important facet of algorithms, that of searching. Searching is a powerful tool that has found many applications in computing. We will look at some of the most common searching algorithms. We will not focus completely on the mechanics of their operation, but instead you will get an understanding of what they are used for and the gist of how they work. Join me as we journey into this chapter on searching.

Breadth-First Search

We will begin our discussion on searching with a look at an important searching algorithm, breadth-first search. In a previous chapter, we looked at the basics of graphs. The breadth-first search algorithm is an algorithm that searches through graphs. Breadth-first search searches all points broadly, beginning at the ones closest to the starting point.

Breadth-first search determines whether there is a path between two nodes and determines what the shortest path between the two nodes is. Let's see how the algorithm works. Ignoring construction specifics, let's imagine we have the graph shown in Figure 8-1.

© Armstrong Subero 2020
A. Subero, *Codeless Data Structures and Algorithms*,
https://doi.org/10.1007/978-1-4842-5725-8_8

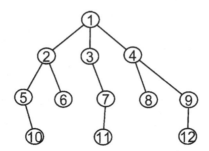

Figure 8-1. Graph

Let's say we want to reach the 10 from the 1. The algorithm can reach the 10 from the 1 through three possible paths, in other words, through the 2, 3, or 4. The algorithm will start at the 2 and then have other possible candidates, namely, the 5 and the 6. Then it will move from the 3 to the 7 and then from the 4 to the 8 and 9. This continues until all the points in the graph have been searched. Afterward, the search will take place starting with the 5, and then it will reach the 10. Once the 10 is reached, the algorithm would have reached its goal, and the search will end.

Breadth-first search essentially searches a graph layer by layer. Figure 8-2 shows us the path that breadth-first search uses to traverse the graph.

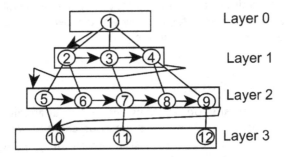

Figure 8-2. Breadth-first search path

Essentially the algorithm starts at layer 0 and then moves horizontally until it visits all the nodes of the layer before it moves to the next layer.

Dijkstra's Algorithm

Have you ever wondered how navigation apps work to find you the fastest route? Or how video data packets find the shortest path on a network for real-time video communication? Problems like these are core to computer science, and while they seem complicated, we can find a solution if we use a graph to represent these problems.

If we set up such problems as points of a graph, we can get a solution to these problems by finding the shortest path between the points on a graph. There are several algorithmic methods we can use to do this; one such method is Dijkstra's algorithm. This algorithm works on a weighted graph. This algorithm tries to find the shortest path from one node to the other nodes in the graph network.

The weights on the graph from Dijkstra's algorithm are called *costs*, and for the algorithm to work, the costs must not be negative. We usually call the starting node on the graph s, and we typically want to find the shortest point from our starting node to other nodes on the network.

Let's imagine we have a graph as in Figure 8-3, and we set the starting s node as node 1. The algorithm will set the value of node 1 to 0, and all the other nodes will be assigned the value of infinity. Nodes 2 and 3 can be reached from node 1, so the algorithm will calculate the distance values for the nodes. For example, for node 2, the value will be 7, and for node 3 the value will be 1. Since node 3 has the smallest distance, it will become the current node. From node 3 we calculate the cost of each candidate associated with the node, in this case, node 4, and then from node 4 this will continue. The algorithm will keep iterating and calculating the shortest path to each point, one path at a time until the last point is reached.

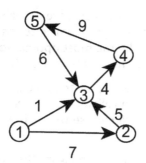

Figure 8-3. Graph for using Dijkstra's algorithm

Dijkstra's algorithm is powerful; however, the limitation with the algorithm is that it consumes a lot of time and resources as it performs what is known as a *blind search* or an uninformed search as the algorithm has no information about the space it is searching. Despite its limitations, however, it finds a lot of uses in a lot of map and navigation algorithms.

There are other algorithms that improve upon Dijkstra's algorithm such as the Bellman-Ford algorithm, which is also used to find the shortest path on a graph. The Bellman-Ford algorithm also searches a graph in which all the edges are weighted. We discussed the point that Dijkstra's algorithm doesn't work for graphs that have negative edges. The Bellman-Ford algorithm, however, works for graphs that have negative edges.

A* Algorithm

With the recent proliferation of interest in robotics and artificial intelligence, there is an important algorithm that you are likely to encounter, the A* (called *A star*) algorithm. Like Dijkstra's algorithm, the A* algorithm finds the shortest path between points.

Before we begin to discuss the A* algorithm, let's discuss the two types of approaches to pathfinding. There is the greedy approach and the heuristic approach. Let's take some time to understand these two approaches.

When we say an algorithm is a greedy algorithm, what we mean is that the algorithm works on a solution to the problem one step at a time, always choosing the path that offers instant gratification, which is to say the path that provides the most immediate benefit to the problem at hand. Think about Dijkstra's algorithm, in that the algorithm selects the current node because it had the lowest value from the previous node. The limitation of greedy algorithms is that they do not consider the problem as a whole and make decisions based only on a single step.

Heuristic algorithms, on the other hand, are probabilistic by nature and try to give an approximate solution to a problem. In terms of pathfinding, a heuristic solution may not necessarily be the most optimal solution, but it will be good enough to work in practice. The A* algorithm falls into this category.

The A* algorithm combines the approach taken by Dijkstra's algorithm with elements of another algorithm called *best-first search* to give a good heuristic solution. The A* algorithm will choose the path with the lowest overall cost of traversal.

Let's imagine we have a situation as in Figure 8-4 and we want to move from the star to the circle as quickly and easily as possible.

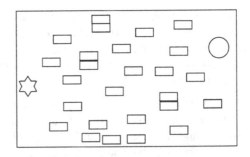

Figure 8-4. Finding the shortest path

To use the A* algorithm, we would first have to consider the problem as a graph. Once we do this, the A* algorithm will calculate the cost of each point to move from the starting location and the estimated cost to the goal. Once this has been determined, the A* algorithm will choose one of the points with the lowest cost. Once we select the point with the lowest cost, we consider that point as "explored." The algorithm then iterates, calculating the cost of each point that can be moved to from the explored point, and the one with the lowest cost is selected and considered explored, and so forth.

Compared to Dijkstra's algorithm, which takes every possible path on a graph, the A* algorithm is more efficient as the heuristic cost of the shortest path can be close to the actual cost of the shortest path. Even if the algorithm fails, you are still guaranteed to find the shortest path with the A* algorithm.

Conclusion

In this chapter, we looked at various searching algorithms including breadth-first search, Dijkstra's algorithm, and the A* algorithm. In the next chapter, we will look at some popular clustering algorithms.

Clustering Algorithms

In this chapter, we will look at clustering algorithms or, as I like to call them, K-algorithms. Specifically, I'm referring to the K-nearest neighbor algorithm and the K-means algorithm. These algorithms are finding extensive use in classification systems and machine learning systems. These algorithms can become complex and mathematical. However, what we will do is look at the principles of these algorithms without the mathematical details of their implementation and simply use the concepts behind the mathematics to explain them.

K-Means Algorithm

The K-means algorithm is an important algorithm that is popular today because of an increased interest in artificial intelligence and machine learning. It is an algorithm that is used in something called *graph clustering*. Graph clustering uses certain properties of the structure of data, for example, heterogeneity, to classify the data. This act of classifying the data based on groups of elements is also called *clustering*.

To use the clustered data, algorithms that perform something known as *cluster analysis* are applied to the data elements. What happens in cluster analysis is that data is grouped based on element descriptions and on the relationships between objects. Essentially, the points within a cluster have different properties.

© Armstrong Subero 2020
A. Subero, *Codeless Data Structures and Algorithms*,
https://doi.org/10.1007/978-1-4842-5725-8_9

To understand the K-means algorithm, we must understand the ways in which we classify types of data clusters. There are two types of data clusters that we must concern ourselves with: hierarchical clustering and partitioning clustering.

Figure 9-1 shows an example of hierarchical clustering, and we can imagine each point in the form of a circle labeled with a letter to be a data element. In practice, most hierarchical diagrams are usually more complex; however, this diagram will suffice for our purposes.

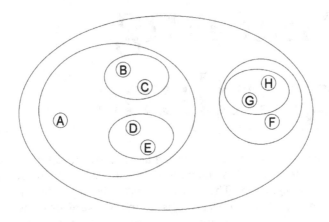

Figure 9-1. Hierarchical clustering

In this diagram, each of the labeled circles is grouped next to each other because they are similar nodes. The nodes within a hierarchical cluster show the relationships between similar sets of data.

Without delving too much into the mathematical operation of hierarchical clustering, all you need to be aware of is that in hierarchical clustering, clusters of data that are close are identified and merged together, and the result is called a *dendrogram*.

Figure 9-2 shows an example of a simple dendrogram that can be used to represent our hierarchical clustering diagram.

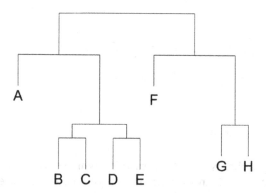

Figure 9-2. Simple dendrogram to represent our cluster

Partitional clustering is the other type of clustering that you will need to know about; if we look at Figure 9-3, you will see what this type of clustering looks like. This is the type of clustering that the K-means algorithm concerns itself with.

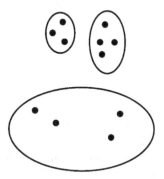

Figure 9-3. Partitional clustering

In the K-means algorithm, within the partitional cluster, there is a center point called the *centroid* that gets points assigned to form a cluster around the closest centroid. Within the programming language (or pseudocode), you must specify the number of clusters designated as "K" that the algorithm will work with. This is the "K" part of the algorithm.

The algorithm usually chooses the centroid randomly, and the close points that are chosen usually take place based on some mathematical property called the *Euclidean distance*.

We can now run through the algorithm step-by-step to get an idea of how it functions. Let's imagine we set the K value as 2 (K = 2), and we have data points on a graph as in Figure 9-4. On this graph we have two centroids; one centroid is the triangle, and the other is the square.

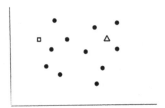

Figure 9-4. Hierarchical clustering

The K-means algorithm will try to determine which data point is closer to the triangle centroid and which data point is closer to the square centroid. The algorithm will determine this with the Euclidian distance by drawing a line on the graph. Everything on the left of the line will be closer to the square centroid, and everything to the right of the line will be closer to the triangle centroid. This is depicted in Figure 9-5.

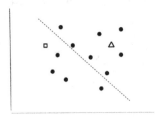

Figure 9-5. Using Euclidean distance

After the data point is assigned to a centroid, the centroid values are usually recomputed based on the data values of the associated data elements, and the average is taken. After the average is found, the centroid is repositioned as in Figure 9-6. This is where the "means" part of the algorithm comes from because to take the mean is to find the average.

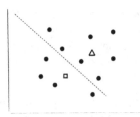

Figure 9-6. Recomputed centroid

This process keeps iterating until no point must move closer to any centroid and the algorithm stops running.

The K-means algorithm finds uses in unsupervised machine learning applications as well as classification applications. In the next section, we will look at the K-nearest neighbor algorithm, which is the other K-algorithm we will discuss in this book.

K-Nearest Neighbor

In the previous section, we looked at the K-means algorithm, so in this section we will look at the K-nearest neighbor (KNN) algorithm, which is a powerful tool in modern computer science applications, particularly machine learning.

The K-nearest neighbor algorithm can get mathematically intensive, so it is best to explain the algorithm with an example. Since the KNN algorithm excels at classification, let's look at a classification example for the algorithm.

Let's imagine we have a classification system that classifies shapes into circles or squares. Now, put these shape data points on a graph and see where they fall according to the shape, as in Figure 9-7.

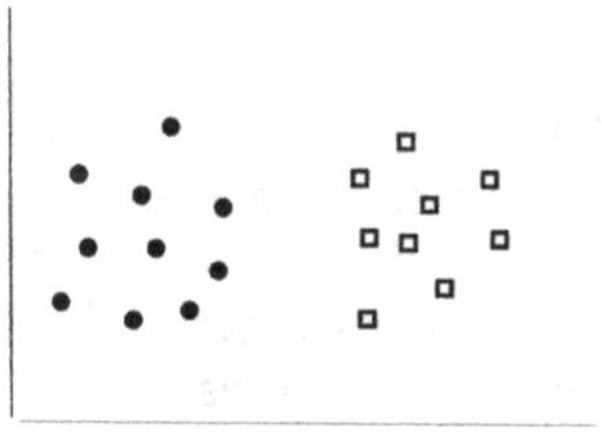

Figure 9-7. Shape classification

Let's say we put a new shape into the mix, like an octagon, as in Figure 9-8.

Figure 9-8. Octagon introduction

Since the octagon is a new shape, how can we determine whether it is a square or a circle? Let's say we set the K value to 5. Within the K-nearest neighbor algorithm, the K values determine the number of nearest neighbors to consider in the classification process.

With a value of K=5, the shape will be assigned based on the number of neighbors it is closest in value to, as in Figure 9-9.

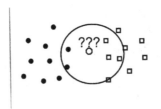

Figure 9-9. Classification

Since the classification process with K=5 determines that the octagon has three square neighbors and two circle neighbors, the octagon will thus be classified as a square since three out of five of the neighbors are square. This is how the KNN algorithm performs classification.

Machine Learning

After our look at the K-algorithms, you may be curious about machine learning. Machine learning is a branch of computer science that ties back into the foundations of the subject from mathematics. This is because machine learning uses statistics to perform its operations.

In its present form, machine learning uses computers as tools that employ statistical methods to analyze data and make intelligible decisions and predictions. To do this machine learning, systems essentially gather a lot of data and convert it into the information that we desire. A machine learning system will take data and analyze it to use it to train machine learning algorithms that will then be used to solve some problems.

Knowing this, we can say that machine learning systems essentially use artificial intelligence to learn from data by being trained with it. It is important to note that artificial intelligence (AI) and machine learning are not the same thing. AI refers to computing systems that can perform intelligent tasks, whereas machine learning systems can learn how to do tasks on their own with limited or no supervision.

When a machine learning system learns from a labeled input to the system, called a *dataset* (the training data), this is known as *supervised learning*. In a supervised learning system, it is expected that the machine learning algorithm will produce a specific answer such as in the classification problem we explored earlier in the KNN section.

When the dataset is not labeled and there is no specific outcome expected from the algorithm, we call this *unsupervised learning*. While this may seem like a waste of time, unsupervised learning allows us to explore patterns in data we did not consider previously or may not have been aware of.

There is also reinforcement learning, where learning is done with feedback to increase performance over time using feedback from previous attempts at performing a task.

Neural Networks

A powerful tool to hit the realm of machine learning is the use of neural networks. A *neural network* is a network of neurons that exist in biological systems. Neurons within biological organisms communicate with each other and are the basic working unit of the brain.

In an effort to increase the abilities of artificial intelligent systems, the artificial neural network (ANN) was devised to mimic the brains of biological organisms. These ANNs are networked to solve problems and, similar to biological brains, have the neurons within the neural network arranged in layers.

Each neuron within the network is connected to every other neuron in the preceding and subsequent layer. The neurons within the layers are called *nodes* or *units*. Figure 9-8 shows us what an ANN may look like.

An ANN typically is comprised of an input layer, one or more hidden layers, and an output layer. The input layer takes in information and introduces it into the network but performs no computation; its primary function is data transfer.

The hidden layer is where the calculations take place. Within the hidden layer of the neural network, computations are performed, and these nodes serve as a bridge from the input layer to the output layer.

Within the output layer, computations may also take place, and these nodes transfer data from within the neural network back into the outside world.

This structure of a neural network shown in Figure 9-10 is known as a *feedforward* neural network based on the flow of data within the network. These feed-forward networks may be called a *single-layer perceptron* if there is no hidden layer or a *multilayer perception* if there is one or more hidden layers.

Input   Hidden   Output

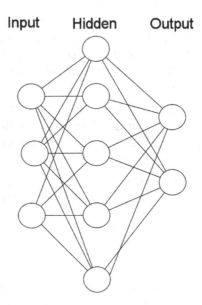

Figure 9-10. An example artificial neural network

Neural networks form the basis of many modern clustering and classification systems and drive artificial intelligence.

Deep Learning

While on its own machine learning is powerful and can be used to perform many analytical and predictive tasks, when it is combined with neural networks, we see the emergence of an even more powerful system.

Deep learning systems are machine learning systems that utilize neural networks to perform machine learning. This allows for a nonlinear learning process as the neural networks are arranged in a hierarchical manner.

Deep learning systems are accurate and allow artificially intelligent systems to sometimes exceed human performance. These systems utilize a lot of data and powerful computing platforms to achieve their results. These deep learning systems have a large depth of layers through which data must pass, which is to say they have more than one hidden layer.

There are many applications of deep learning in everyday life. Voice recognition systems, self-driving vehicles, and medical research are just a few examples.

Conclusion

In this chapter, we looked at two of the K-algorithms: K-nearest neighbor and K-means. We also briefly delved into machine learning and neural networks.

With the completion of this chapter, we have covered the core algorithms you will meet as a developer or student. These algorithms will serve you well for the lifetime of your computer science career. In the next chapter, we will look at a topic that isn't part of DSA per se but is a subject that I think belongs in a DSA book as it will help your understanding of some aspects of DSA. We will look at the concept of randomness.

Randomness

We are almost at the end of our journey. Assuming you have gone through all the chapters in this book, you will have seen many data structures and algorithms that you will encounter as a developer, builder, programmer, or whatever term you use to describe yourself. From this chapter onward, we will cover additional topics in DSA. Of course, there are more topics you may encounter, but with the foundation of knowledge you currently have, you should be able to understand them. In this chapter, we will briefly discuss randomness in computers. This topic is useful because often when designing your own algorithms, you may need to use random numbers for some purpose. There are even randomized algorithms that utilize the concept of randomness for their operation. This chapter helps you understand what is going on under the hood.

Random

The word *random* is one that we use regularly. When we say something is *random*, we mean it occurs without a discernible pattern and involves unpredictability. Humans have an intuitive concept of randomness. Humans can be very unpredictable. Though some people have what those in law enforcement call a *modus operandi* (essentially a working habit or a way they operate), many people are able to act spontaneously or on impulse. The human mind is not easy to predict, and even the best psychologist cannot explain all human behavior.

© Armstrong Subero 2020
A. Subero, *Codeless Data Structures and Algorithms*,
https://doi.org/10.1007/978-1-4842-5725-8_10

Computers, on the other hand, by their very nature do not have the same innate ability to be random. Computers must be programmed to give a particular output, and even if it seems random, it's actually based on some algorithm. This is because they are designed to execute the exact instructions that are given to them, which makes them very predictable. A popular saying within the realm of computing is *garbage in garbage out* (GIGO).

Since the extent of randomness can span many domains, we will restrict our discussion to random numbers as they are the most easily understood and are most widely used in the programs that you may have to write.

It is hard to discuss random numbers without looking at some complex mathematics. However, we'll skip the hard math part and replace it with English and good descriptions.

In mathematics, there is a concept called *probability*. Probability is the chance of something occurring. If we have a coin and flip it, there is a chance it will land on heads, and there is a chance it will land on tails.

Now what would happen if you flipped a coin more than 100 times? What would the outcome be? How many heads? How many tails?

If you ask most people how they think the coin flip distribution will look, they may tell you that the coins will have some sequence of heads and tails that cannot be determined. However, there is a chance it can be all heads or all tails.

This is the essence of the concept of randomness; random things can take a form or pattern that we cannot expect. When something is random, whether it be coins or numbers, it can take any form of distribution.

To bring some order to the world of randomity, mathematicians have come up with the concept of probability to analyze randomness; however, even probability cannot truly predict randomity.

Some Hardware Background

To fully understand some of the topics within random number generation and understand how random number generators work, it is important to gain an understanding of what occurs in the low-level hardware of your computer.

When studying topics in computer science, sometimes we become so abstracted away from the hardware that we forget a computer is an electronic device. Behind all those algorithms, data structures, and software frameworks is the underlying hardware. Throughout this book, as we discussed data structures and algorithms, efficiency was one of the central pillars of this topic. To have a solid grasp of efficiency, you must be able to understand what your code is doing at the fundamental level, which is manipulating hardware.

In the succeeding sections, we will look at some of these concepts of the underlying computer hardware without any of the intricate details. If you understand these topics, you can become a better algorithm designer. Writing efficient algorithms is akin to being a good driver, and when you understand the hardware, you transition from being simply a driver to knowing some of the mechanics of the vehicle. Knowing the mechanics of the vehicle will no doubt make you a better driver.

Circuits and the Transistor

To begin our discussion of hardware, we will start with the transistor. The technical revolution of classical computing started with the transistor. The transistor is the fundamental component of the electronic building blocks of your computer. Your CPU with all its internal logic, as well as your computer memory, is made up of transistors in various configurations.

A discrete transistor is a three-terminal device that can be used for switching electronic signals and electrical power. Within the realm of electronics, we use something called a *schematic diagram* to represent the various interconnections of electronic components. Figure 10-1 shows an example of a schematic diagram.

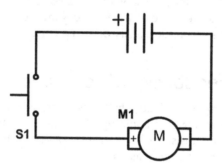

Figure 10-1. A schematic diagram

In this schematic diagram, the alternating long and short lines with a plus sign represents a battery. The circle with an M in the center represents a motor, and the sideways T symbol over a space represents a switch. If you look carefully, you will also notice that each of the components is labeled. The switch is labeled S1, and the motor is labeled M1.

You will see the connections between each of the circuit elements denoted with a line. This line represents the wires that are connecting everything together.

Figure 10-2 shows the schematic symbol for a transistor.

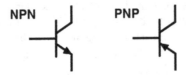

Figure 10-2. Schematic symbol of a transistor. In the didagram NPN stands for Negative Positive Negative and PNP stands for Positive Negative Positive. These "Positive" and "Negative" refer to the p-type or n-type semiconductor materials that contain a substance known as a dopant that give the device it's properties.

In modern electronic devices, there is a more efficient transistor that has taken the original transistor's place. We call this more efficient transistor a *metal-oxide semiconductor field effect transistor* (MOSFET). When you connect MOSFETs together, you can create your memory and CPU and consume much less power. For our purposes, though, when we say *transistor*, we may mean the actual transistor or a MOSFET; for our learning's sake, they are interchangeable.

Transistors are so important to computing that there is a direct correlation between how many transistors you can pack on a chip and how powerful your computer is. We call this Moore's law. Moore's law states that we can expect our computers to become more powerful every few years, and this is because of our ability to make and pack transistors on a chip.

Amplifiers, Feedback, Clocks, and Oscillators

In the previous section, we talked a little about transistors, and we said that they are used for switching electronic signals within circuits. There is another purpose, though, that transistors have. Transistors have the ability to perform amplification of electronic signals as well.

When we speak about amplification, what we are talking about is the ability of a circuit component to take a weak signal and produce a more powerful signal from it. We measure the ability of a component to perform amplification in terms of gain.

When we talk about gain, we mean the ratio of output voltage (electrical pressure), current (rate of flow of electrons in a circuit), or power (a combination of voltage and current) to the input of the component performing the amplification.

In terms of amplification ability, the transistor is pretty good; however, when we combine the transistors together with other electronic components, we get a device called an *operational amplifier*. The schematic symbol of the operational amplifier is given in Figure 10-3.

Figure 10-3. Schematic symbol of an op amp

The op amp uses *feedback* to operate. Feedback is the name given to the process of taking some of the output of the device and feeding it back into the input. There is also negative feedback, which feeds some of the output back into the device but into the inverting terminal of the device. A *terminal* is the name given to the pins of the device. In the schematic symbol in Figure 10-3, the + terminal is the *noninverting* input, and the - terminal is the *inverting* input.

One function we can use an op amp for is a *comparator*, which is a device that is used for comparing two electrical signals and then doing something on its output. A comparator works by having a reference voltage on one of the pins. If the other pin is above or below that reference voltage, the comparator will produce a logical HIGH or a logical LOW on its output.

When talking about CPUs, something that comes up quite frequently is the mention of a clock. We always hear that something can be done in a number of clock cycles or the CPU runs at a particular clock speed. Even when we design our algorithms, we want them to be as efficient as possible, which means that we want them to occupy as minimal a clock cycle time as possible during execution. Clocks are like the heartbeat of digital circuits.

A clock is a type of signal that varies between high and low at a given interval. This variation between high and low is known as *oscillation*. There are special circuit configurations known as *oscillators* that produce a waveform periodically, and such circuits are used to produce the clock that drives digital components such as your CPU and memory.

Logic Gates

Transistors can be combined to form what is known as *logic gates*. A logic gate is a device that gives a particular output of high or low depending on the state of the inputs, which can be either high or low. Figure 10-4 shows some of the important logic gates.

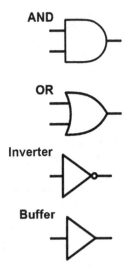

Figure 10-4. Schematic symbols of various logic gates

Starting at the top, we have the AND gate. If you look at the AND gate, you will see two terminals on the input side and one terminal on the output side. The output terminal of the AND gate produces a logical 1 or HIGH if the two inputs of the gate are HIGH, and a logical 0 or LOW if only one input is HIGH while the other is LOW or the two inputs are LOW.

The OR gate will produce a logical 1 or HIGH at its output if either of the two input terminals are HIGH or both are HIGH. If both the input terminals have a LOW state, then the OR gate will produce a logical LOW.

The inverter also known as the NOT gate will produce the opposite of what the input is. If the input is LOW, the output will be high, and if the input is HIGH, then the output will be LOW.

At the bottom is the buffer gate, which simply passes whatever is at the input. If the input is HIGH, the output is HIGH, and if the input is LOW, the output is LOW. This gate is used for translating one logic level to the next and for providing logical isolation whenever it is needed.

Another gate you are likely to encounter is the XOR gate. The XOR gate, also known as the *exclusive OR gate*, is a gate that produces a high output when the values at the input gates are odd. When both inputs are HIGH or LOW, the output of the gate is LOW. When one input is HIGH, the other is LOW. Figure 10-5 depicts the schematic symbol for the XOR gate.

Figure 10-5. The XOR gate

The XOR gate is one type of Boolean function anyone involved in cybersecurity is likely to encounter. This gate is used in parity detection circuits and heavily used in cryptographic functions. It can even be used in certain types or random number generator circuits, as we will discuss later in this chapter.

The logic gates provided here are just some of the many, many gates that exist and are used in the construction of your digital computer. Keep these in mind as we will look at them again when we discuss how your computer generates random numbers in hardware.

Combinational and Sequential Logic

When you combine logic gates, you get what is known as *combinational logic* or *sequential logic*. Combinational logic circuits are circuits whose outputs are determined by the logical function of their input state. Sequential circuits, on the other hand, implement memory and feedback loops to have circuits that not only take their present state into account but also utilize previous states to produce their output.

Sequential circuits usually utilize clocks, pulses, or some events to operate. Within the sequential logic circuits, you should know about the flip-flop. The flip-flop, also known as a *latch*, can be used to store information about its state. Flip-flops form the basic building blocks of computer memory.

In Figure 10-6 we see one of the simplest types of flip-flops, the D flip-flop. The simplest D flip-flop has a data input D and a clock input (represented by the terminal leading to a triangle on the body).

The output Q is dependent on which signals are applied to the inputs. Like we already mentioned, a flip-flop is one of the basic units of memory, and we call the basic unit of storage for a CPU a *register*. So, a flip-flop can be thought of as a single register.

D Flip-Flop

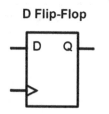

Figure 10-6. The D flip-flop

There are many different types of flip-flops, and when we combine these flip-flops, we can get what is known as a *shift register*. A shift register is a circuit that comprises many flip-flops that store multiple bits of data. Figure 10-7 shows us an example of a simple shift register, specifically a 3-bit shift register.

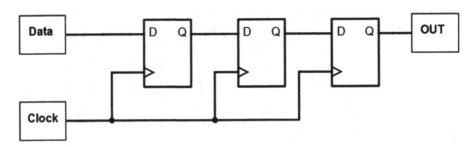

Figure 10-7. A 3-bit shift register

In a shift register, you can input bits into the system either serially or in a parallel manner. There are *serial in serial out* (SISO) shift registers that take data in serially and output it serially. *Serial in parallel out* (SIPO) shift registers are where data is input serially and then output all at once in parallel form. There are also *parallel in serial out* (PISO) and *parallel in parallel out* (PIPO) shift registers.

Shift registers can be combined with other circuits to perform some interesting bit manipulation functions.

Mixed-Signal Circuits, Reactance, and Noise

The circuits we have discussed so far are either digital (gates, flip-flops, shift register) or analog (op amps). However, one circuit we discussed is somewhere between the two. We said that the comparator compares two signals at its input and then does something at its output.

A comparator takes a continuous signal (analog) and converts it into a digital one (discrete). The process of converting an analog signal into a digital one is called *analog-to-digital conversion* (ADC). Similarly, you can convert digital signals into an analog one, which is called *digital-to-analog conversion* (DAC). These conversions are important because our world is analog in nature.

A good example to explain this is in the recording and playback of digital sound. If you record sound through a microphone, there is a circuit that will take the continuous signals being produced by your voice and convert them into discrete digital signals so that the CPU can process and store this information.

These circuits that contain both analog and digital components are known as *mixed-signal circuits*. As we will see later when applied to random number generation, these circuits are important when we look at true random number generators later in the book.

When you play back the sound, there is another circuit that will take the digital signals being provided by the CPU and convert them into a continuous signal representation so that you will be able to hear the sound that was recorded.

Within the realm of computer hardware, there are certain circuit elements known as *capacitors* and *inductors*. These circuit components exhibit *reactance*, which is the act of resisting the flow of alternating current (AC). AC current alternates or oscillates.

Sometimes in our circuits there may be the existence of noise. Noise is the name given to an unwanted disturbance in an electronic signal. Sometimes you may also hear about harmonics, which are unwanted higher frequencies in an AC signal.

Now that you have a basic understanding of the underlying computer hardware, we can look at the concept of randomness in computing circuits.

Pseudorandom Numbers

Experienced programmers may say that computers do have the ability to be random and may point out that X language does provide facilities for random number generation. In fact, one such programmer I spoke to, let's call him R, never quite understood what I was explaining to him about random number generation. R was a great self-taught programmer, but R never spent time learning any theory. As a result, most computer science topics to him were "worthless," and he never really cared to learn more; little does he know how much he is missing out on. I'll now have the same discussion with you that I had with R about pseudorandomness.

Before we delve into all the complexities of what being pseudorandom means, let's take some time and discuss what it means to be *pseudo*. After all, we already discussed what it means to be random, so we need to discuss pseudo.

My first encounter with the word *pseudo* was in fact not in computer science but within chemistry. A scientist by the name of Dmitri Mendeleev predicted the existence of a chemical element that he called *eka-silicon* because he knew it would be "like silicon" though it wasn't silicon. (This eka-silicon was later discovered and named Germanium.) While reading about the concept of "eka" elements, I ended up somehow learning about alchemy, which is a pseudoscience that existed around medieval times.

You see, alchemists were obsessed with turning other elements into gold, but the transmutation of elements was not possible with the knowledge and technology available at the time. While alchemy paved the way for the science of chemistry, the art in and of itself is not science.

Like alchemy is to chemistry, the word *pseudo* refers to something that appears to be one thing but is not.

Within computing, many programming languages provide facilities for generating numbers that appear to be random but because of the design of the algorithm responsible for their generation are not actually so. We call these numbers *pseudorandom numbers*, because while they appear to be random numbers, they are not random.

These pseudorandom numbers use something known as a *seed number* (provided by the programmer), which is then used by an algorithm to generate a sequence of numbers. While from the perspective of the user of the program the numbers will appear to be random, upon analysis of the numbers generated, one will see that the same sequence of numbers is repeated.

With pseudorandom number generation, if the seed is the same, then the generated sequence of numbers will be the same. To combat this problem, many programmers will select something that is changing (usually some parameter provided by the operating system) and then run it through a hash. Such a sequence of operations will generate something that is close to a true random number.

Though pseudorandom number generators are not real random number generators, in practice whenever you hear engineers talk about a random number generator (RNG), they are really talking about a pseudorandom number generator (PRNG).

Pseudorandom numbers have their place in places like video games and other noncritical applications. However, for applications like cybersecurity, particularity encryption, if we use pseudorandom numbers, then they will be able to be guessed by the attacker.

There are special PRNGs designed for cybersecurity applications. We call these PRNGs *cryptographically secure pseudorandom number generators* (CSPRNGs), which are also known as *cryptographic pseudorandom number generators* (CPRNGs).

These PRNGs are complex and exist in different variations intended for different security applications. As good as CSPRNGs are, there are limitations to such types of number generators. For such applications a PRNG is not good enough, and we need to look at a different type of random number generator that defies the very nature of computing systems, the true random number generator.

Linear-Feedback Shift Register

One method of generating pseudorandom numbers is the utilization of a linear feedback shift register. A *linear feedback shift register* (LFSR) is a device that uses the combination of flip-flops with a XOR gate to generate a pseudorandom binary sequence. A simplified example of this circuit is shown in Figure 10-8.

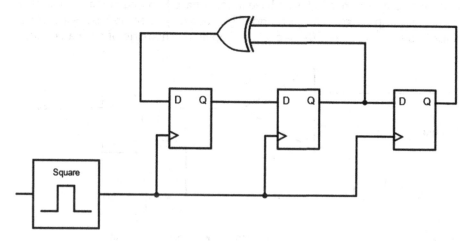

Figure 10-8. A basic linear feedback shift register

The initial seed value we give to the LFSR will enable it to generate a sequence of bits that appear to be random in nature. When you feed a number into the LFSR, the output is given as a linear function of the input. Because of the nature of the LFSR, the pattern of bits that are produced will repeat periodically; however, since the maximum length sequence of this pseudorandom number generator binary sequence is long, it can make a good pseudorandom number generator for noncryptographic purposes.

The LFSR-based pseudorandom number generator is popular because you can implement both hardware and software-based implementations of this generator. It is simple and effective, and a software-based implementation of this pseudorandom number generator can be run on resource-constrained systems such as 8-bit microcontrollers.

True Random Number Generator

A defining characteristic of a PRNG is that they are dependent on some software algorithm or rather simplistic hardware. The TRNG, however, typically uses a more complex hardware random number generator (HRNG) to generate their random numbers, such as some random event from noise or some chaotic system.

For example, a HRNG system that I commonly encounter is based on an oscillator. We can use a logic system to generate a free-running oscillator and add some negative feedback to it, and then we can add a reactive circuit to generate some stray reactance (essentially noise). We can use other logic circuits to detect this noise and use it to generate a random number.

Another common method is to have a noise-based source and a comparator with a flip-flop to generate a random bit that may then be read via a clock pulse request. In Figure 10-9 we see a simplified schematic of such a circuit.

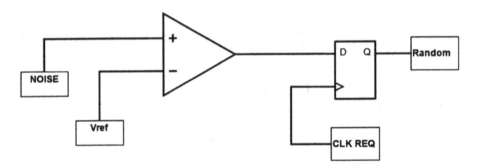

Figure 10-9. A basic hardware noise-based true random number generator

In this circuit we have a reference voltage to set the bias point for the comparator. If the noise level voltage is below the reference voltage of the comparator, the output is a logical LOW or zero. If the noise level voltage is above the reference voltage, then the output will be a logical one.

There are many such systems to generate random numbers that may use some noisy systems, oscillation-based systems, and even some quantum-based systems for generating true random numbers.

With the prevalence and recent hype of connecting embedded systems to the Internet, which we call the Internet of Things (IoT), there has been a need to provide data security. As such, some microcontroller and microprocessor manufacturers have started incorporating TRNGs into their microcontrollers. As technology expands and we demand an Internet of Everything (IoE), in which our cars, homes, factories, cities, and even the plants in our home are connected to the Internet, we will need more secure devices connected to the Internet. RNGs, especially TRNGs, will have a pivotal role in that future.

Conclusion

In this chapter, we briefly touched on the concept of randomness, looked at some low-level computing concepts, and spoke about pseudorandom numbers and true random number generation. In the next chapter, we will look at algorithms used in operating systems scheduling.

Scheduling Algorithms

We have reached a point in our discussion where we have looked at most of the common algorithms that are in use today. However, our discussion has not yet finished as we can look at some of the less talked about algorithms, the algorithms that run without our operating system. In this chapter, we will explore some of the concepts of operating systems and look at some of the common scheduling algorithms that run within operating systems. The knowledge provided here can help you gain a deeper understanding of how some of the data structures you learned earlier can be implemented to facilitate the execution of algorithms. As you will see, much of the data structure knowledge you learned before will be used here within the implementation and organization of the scheduling algorithms.

Operating Systems

The world of computing as we know it today is dependent on operating systems (OSs). Supercomputers run some variant of Linux, home computers run macOS and Windows, and our mobile computing platforms are dominated by Android, which is, of course, based on Linux.

Our modern world of connectivity and information availability is possible thanks in part to a lot of people working for and perfecting the operating systems that run our computers. However, have you ever considered what operating systems are and why they are necessary?

© Armstrong Subero 2020
A. Subero, *Codeless Data Structures and Algorithms*,
https://doi.org/10.1007/978-1-4842-5725-8_11

Picture this scenario: you are working on a paper or document in your word processor of choice while listening to music. While you are doing this, you get a notification from your social media account and another notification that a new video has been uploaded. You load your web browser to check your social media and then open a new tab to look at the video. You leave both tabs open and go back to work while listening to music.

These tasks seem trivial and are tasks users expect of modern computing platforms. Multitasking is commonplace in today's computers and is something that we take for granted.

You see, our current classical computing CPUs have finite resources. Memory and processing time must be shared among the many programs that are running on the system. The operating systems provide a mechanism for abstraction to allow programs to run simultaneously; your media player, web browser, and word processing document all run simultaneously and use the same processor and resources of your computer, thanks to your operating system.

Operating systems run not only on servers, laptops, and mobile phones, but also on the tiny microcontrollers within the embedded systems in our photocopiers, cars, toys, and gadgets around us. This brings us to our discussion of the two types of operating systems, general-purpose operating systems and real-time operating systems.

General-Purpose Operating Systems

A general-purpose operating system (GPOS) is an operating system that is designed for general computing tasks and applications. GPOSs run on standard computer hardware where memory is plentiful, and the user may run one or more applications at any given point in time.

Most general-purpose operating systems are designed to run on desktops and laptops. Windows, macOS, and heavyweight Linux flavors such as Ubuntu and Linux Mint fall into this category. GPOSs also target workstations, mainframes, and embedded devices that have the processing and memory requirements necessary to run the system.

With the increase in computing power on mobile devices, mobile operating systems like Android and iOS can be classified into this category.

Real-Time Operating Systems

When you are using a general-purpose operating system, you may be accustomed to having programs on your computer or phone crash, freeze, restart, or run slowly from time to time. If you are using a web browser, for example, and you see a message stating the program crashed unexpectedly, then it's not really a big deal; you simply close the window and reopen the browser. If you are using an application and click a button and the computer delays a while before responding to your input, you will be annoyed; however, besides mild annoyance, you won't really encounter any problems.

There are systems, however, where these scenarios could be catastrophic. Think of a pacemaker or an autopilot guidance system. In such systems there is no margin for time lapse, delayed response, or crashes. Such systems must be deterministic and respond to events in a timely manner.

We call such systems *real-time systems*. There are *hard* real-time systems that must meet the event response times without any flexibility because failure to do so may result in serious damage to property or loss of life. Think of an ABS in a car or a cardiopulmonary life support system. There are also *soft* real-time systems that must also meet strict deadlines, but there is some degree of flexibility involved. Think of telephony or real-time video conferencing software. In such cases, failure to perform the processing within a specific time may result in a diminished experience for users of the system; however, it will not result in loss of life or damage to property.

Real-time operating systems (RTOSs) are responsible for managing such systems and ensuring that they meet the required processing time requirements.

When compared to general-purpose operating systems, real-time operating systems typically have faster performance and reduced memory requirements in order to meet the real-time requirements. Real-time operating systems must also be reliable to operate for long periods of time without external intervention.

A good example of this would be something like a planetary rover for which it is impossible to have a human repair or reset the device if there is a failure. It is in such applications that RTOSs shine because they provide the reliability needed for such systems.

Some real-time kernels where all the functionality of the operating system takes place within the kernel itself is known as a *monolithic kernel*, which may itself be structured or unstructured.

Interrupts and Interrupt Service Routines

We should talk about interrupts and interrupt service routines before going further. Let's imagine you are working at your desk busily completing a task. Your boss walks up to you with some paperwork. In such a scenario, you would immediately stop what you are doing, take the paperwork from your boss, listen to the instructions, and then go back to working at your desk.

Within the realm of computing, the name we give to such a process is called the *interrupt*. Interrupts tell your CPU that there is something that needs immediate attention. The processor would leave what it is doing and service the interrupt before resuming its task.

The block of instructions the processor must execute when it enters an interrupt is known as the *interrupt service routine* (ISR).

Interrupts may be either hardware-based or software-based. Hardware interrupts are generated by some external hardware device (switch, mouse, and keyboard, for example), whereas software-based interrupts arise from some instruction or program exception that would cause the processor to enter an ISR.

Interrupts have what is known as *interrupt priorities*, which is to say some interrupts are more important than others. Different CPUs have different interrupt mechanisms; however, generally, if one interrupt has a higher priority than another one and they are both triggered at the same time, the interrupt with the higher priority will be serviced before the interrupt with the lower priority.

Finite-State Machines

In getting deeper into our discussion on operating systems, we will look at one of the central pillars of computer science, the finite-state machine (FSM). FSMs, also called *finite-state automata* (FSAs), are used for modeling computation systems that can have changes in state. FSM systems have a finite number of states, and the flow of logic execution is dependent on the flow or transition between states. The transition from one state to another depends on the current state the machine is in, as well as the input into a state.

FSMs can become very mathematical and complex; however, we will look at an example of the FSMs in action to get an idea of how they work.

The simplest FSM that is simple to understand is the coin-operated turnstile that is common in vending machines, gumball machines, and other places where coins or tokens are accepted. This is illustrated in Figure 11-1.

The coin-operated turnstile finite-state machine has two states, locked and unlocked. This FSM works as follows: if you insert a coin into the system, the turnstile will be unlocked. After you push or turn it, it locks again. If the turnstile is locked, then you will not be able to change its state by simply pushing on it or turning it; you must insert a coin. Similarly, if you already inserted a coin and the turnstile is unlocked but you did not push or turn it, then the turnstile will remain unlocked.

The coin-operated turnstile FSM depicted in Figure 11-1 is known as a *state diagram*. Within a state diagram you have nodes, which are the circles. There are also transitional arcs on a state diagram that represent specific conditions that, when assessed as true, will cause the logical flow to take the path of the transitional arc. These transitional arcs are responsible for connecting states to one another. The arcs are labeled with the event that caused the transition.

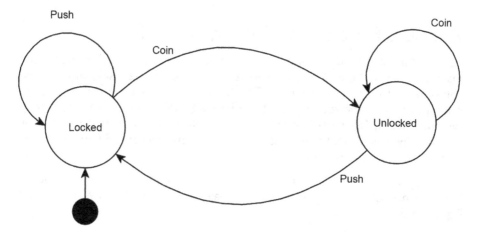

Figure 11-1. Coin-operated turnstile FSM

The black dot pictured in Figure 12-1 is known as the *initial pseudostate*, and there is also sometimes a final state that is depicted by two concentric circles with a solid filled circle surrounded by a circle outline. This is shown in Figure 11-2.

Figure 11-2. Final state concentric circle representation

FSMs typically have some activity that can cause entry into the state, an activity that is performed while in the state, and some activity that can be performed while leaving the state. FSMs can also be deterministic or nondeterministic. In a deterministic state machine, also called a *deterministic finite automaton* (DFA), a given state will undergo one transition given a possible input. In a nondeterministic system also called a *nondeterministic finite automaton* (NFA), a given state can undergo one, multiple, or no transitions when given an input.

In addition to state diagrams, FSMs may also be represented by algorithmic state charts (ASM charts) usually used in digital design and state transition tables. This section provided a brief overview of FSMs. Keep this information handy moving forward.

Kernels, Processes, Threads, and Tasks

At the heart of most operating systems is the kernel. The kernel resides in the main memory and mediates all access to system resources, especially processor time and memory resources. The kernel in of itself can be used as an operating system for your device. In embedded applications, many "operating systems" are in fact real-time kernels. When you add file management, user interface, protocol stacks, and security mechanisms, to name a few, then your kernel becomes an operating system.

When a program is loaded into memory and executed, we call that a *process*. When a process is being executed, it may be in one of many states. See Figure 11-3.

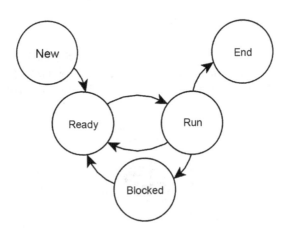

Figure 11-3. States of a process

The process is in the new state when it is first created. When in the ready state, the processing is waiting to be given processor time. Once the process is allowed CPU time, the process enters the run state and can execute instructions. A process can also enter a blocked or waiting state if it requires some resource to continue execution. Finally, after execution, the process will enter the end or exit state and is removed from the main memory.

Sometimes some processes may require several threads to aid execution efficiency as threads allow for parallelism for the execution of a process. If the process is simple, it may require a single thread, and threads are sometimes called *lightweight processes*.

Processes can be classified as either an I/O-bound process, which spends its time completing input and output processes, or a CPU-bound process, which is dependent on the CPU speed for its execution.

Sometimes you may also hear the work *task* being used. A task can refer to part of a thread or part of a process, and the word itself will depend on the operating system being discussed. However, when we talk about a task, what we are usually referring to is just a piece of code or program instruction that is schedulable by the scheduler.

Memory Management Unit

All this talk about kernels and processes leads us to discuss a little about a part of the processor known as the *memory management unit* (MMU). The memory management unit in its rudimentary form provides a mapping between the virtual and physical addresses. When you think about how a CPU accesses memory, most simplified diagrams will show something like Figure 11-4.

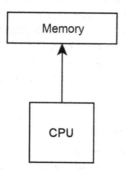

Figure 11-4. How we think the CPU interacts with memory

While this is a good model for simple explanations, there is a middleman between the CPU and memory, and that middleman is the MMU, as shown in Figure 11-5.

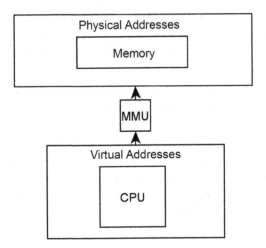

Figure 11-5. How the CPU interacts with memory through MMU

Additionally, paging occurs between memory and the secondary storage disk, which is something to keep in mind when you think about memory. One of the functions of the MMU is that it interacts with the page table to determine what can be done with bits in memory, which is to say whether they can be read, written, or executed.

With the kernel, page table, and MMU working in tandem, processes can then be allowed to have two sets of types based on the execution mode of the process. There is the kernel mode that has access to the system hardware and can access address space for both the user and the kernel. There is also the user mode, which has no direct access to the underlying system hardware and cannot access memory outside of the user address space.

Task Control Blocks

The kernel performs a lot of management of processes. It handles communication between processes, manages processes, and synchronizes processes. The kernel can also handle context switching, which essentially allows the CPU to switch from one process to another without any conflict.

The kernel also creates a *task control block* (TCB), which contains information the kernel needs to know about a task. It may contain things like a process ID or priority info on a process, for example.

The Scheduler and Scheduling

Within any operating system there exists a process scheduler. The scheduler is responsible for determining which tasks execute when. When the scheduler switches from one task to another, we call that a context switch. The part of the process scheduler that performs the context switching and changes the flow of the program execution is called the *dispatcher*.

A scheduling queue refers to the queues of processes. There are generally three types of queue types in a scheduling queue, which are the job queue, the ready queue, and the device queue. The job queue lists all the processes that are awaiting main memory allocation. The ready queue has all the processes that are residing in main memory and are waiting to be scheduled to get CPU time. The device queue refers to the set of processes waiting for allotment to input/output devices.

When we talk about scheduling, there are two types of scheduling: preemptive scheduling and nonpreemptive scheduling. In preemptive scheduling, tasks have priorities that allow tasks with higher priorities to interrupt those with lower-level priorities that are executing on the CPU. There is also nonpreemptive scheduling in which once a task is assigned to the CPU, it keeps running until it terminates or switches out.

To determine which task to execute at which point in time, there are several algorithms that were devised to allow for efficient sharing of the CPU resources. In the following sections, we will look at some of the most common scheduling algorithms.

CPU scheduling algorithms are important because some processes utilize a lot of CPU time. It is important that the scheduling algorithm be able to properly manage the processes in such a way that very time-intensive processes will not utilize all the CPU time.

There are many different methods the scheduler uses to balance CPU time being shared among tasks, which we will look at in the upcoming sections.

First Come First Serve

I want you to envision a scenario with me. Picture you are the owner of a supermarket that has recently opened. You have a large store with eight cashiers. All the customers, whether they have one item or 100, must line up and wait to be serviced by any of the available cashiers. You have the lowest prices in town, so you have no trouble getting customers. From the first day you opened, people come in and fill their carts with items.

As the owner, you see no problem with the current setup. In your mind, customers come, take their items, and proceed to the cashier. After they reach the cashier, they take their goods and leave the supermarket.

Now picture that you are a jogger who just finished your 10k run. You are very thirsty and you ran out of water, so you go into the first grocery store you come to. You grab a bottle of water and proceed to the checkout. There are eight cashiers; however, all currently have a line with at least four carts, and each cart has a lot of items. You want the water, but you are too thirsty to wait in line. Reluctantly, you put the water back on the shelf and go back to your run, thinking that you will stop at the next shop you see.

This scenario may not seem like it has anything to do with DSA; however, if we imagine the eight cashiers as the cores of a CPU and the shoppers as processes, you will begin to see the similarities. In this scenario, customers with carts can be thought of as CPU-intensive processes, and the jogger can be thought of as a lightweight process.

The first scheduling algorithm we will look at is the *first come first serve* (FCFS) algorithm. This algorithm behaves just like the checkout process in this scenario. In Chapter 2, we looked at the first in first out (FIFO) data structure. The FCFS scheduling algorithm behaves like a FIFO.

In Figure 11-6 we are reminded of what the FIFO queue looks like so that we can envision how the FCFS scheduling will work.

Figure 11-6. A FIFO queue

In the FCFS scheduling algorithm, whichever process arrives first gets executed first. The FCFS algorithm can use the queue data structure for scheduling processes; in this arrangement, a new process will enter the tail of the queue, and the scheduler will select a process from the head of the queue.

The TCFS scheduling algorithm is not really the best algorithm to implement for scheduling as heavyweight processes will occupy all the CPU time, leaving little room for other processes to execute.

This FCFS scheduling algorithm will result in a diminished user experience. This is because if heavyweight tasks or processes are running in the background, the response time to user input, which may be a task or process in another program, may never occur. When this happens and the program appears unresponsive to the user, the user experiences a system "lockup."

This type of scheduling algorithm can also lead to total system crashes; this is because this scheduling algorithm does not take priority into account, and if a low-priority task is CPU time intensive, it may prevent critical system tasks from accessing the CPU, and this will lead to total system failure. This is known as the *convoy effect* where the entire system slows down due to one process hogging all the system resources.

There are some instances, for example, in resource-constrained systems, where such a scheduling algorithm may be utilized. However, this type of scheduling algorithm is avoided as there are much better scheduling algorithms, as we will see in the subsequent sections.

Shortest Job First

As the owner of the supermarket, you notice that even though you have the lowest prices in town, your customer turnover is high. Customers seem excited when they come in but leave disgruntled, and you even notice that some of the customers walk into the grocery store, pick up a few items, but then after waiting in line, put the items back on the shelf or abandon their carts and leave.

You do some interviews with customers and realize that many of the customers with a few items cannot get access to a cashier when they come to pay for their goods. This is because they must wait for the customers with filled carts to finish paying before they can pay for their items.

The solution you come up with to the problem is to implement a policy in your supermarket that the customers who have fewer than 10 items can skip directly to the head of the line. Once they have been serviced, then the cashier can resume servicing the customers with carts that are filled.

This is the premise of the second type of scheduling algorithm we will look at, which is the *shortest job first* (SJF) scheduling algorithm. In the shortest job first algorithm, the scheduler assigns processes that have the shortest duration to the CPU to be executed.

Within the SJF algorithm, there are two types. There is preemptive SJF scheduling and nonpreemptive SJF scheduling. In preemptive SJF, as the processes arrive into the scheduler's ready-state queue, if a process is executing and a process with a shorter duration of execution arrives, the existing process is preempted, and the process with the shorter time can run.

In the nonpreemptive SJF, when processes arrive in the ready-state queue, the scheduler will allow the processes with the shortest duration to execute first, then the next shortest duration, and so forth. However, since not all processes arrive at the same time, if a process with an even shorter execution time arrives, it will not be able to run as it will have to wait for the longer process to finish; if the process that is currently executing is very long, then the process will experience starvation and will not be able to execute until the longer process is finished.

While this method of scheduling greatly reduces the average process waiting time over the FCFS method, it is not without fault as for this method to work, the scheduler must have an estimate of how long the task will take to execute on the CPU. Each job, therefore, has an associated unit of time to complete. The prediction of how long a job will take, however, is a hard one and is not always accurate. You may be able to solve this problem by recoding how long the process takes to execute for future reference for the processor; however, this will understandably add some processor overhead.

So, while this method is better than the FCFS method, it is not the best. After all, how would you feel having a cart full of items waiting to pay and seeing a guy with one item zip in front of you only to see him grabbing one item after another off the shelf and carrying on a long conversation with the cashier. So, while it was predicted that he had only one item to pay for and would take a short time, in reality he ended up taking much more time due to other extenuating circumstances.

Priority

Going back to our supermarket scenario, while your previously existing policies worked well, you decided to implement a new brilliant policy that you came up with. You come up with a scheme that will not only make you more profits but allow customers who are always in a hurry to bed satisfied.

Instead of interrupting customers with carts who are paying, you decide to have a premium checkout policy where customers who are in a hurry can purchase a card that allows them to go to the front of the line as long as there are no other card holders. You also offer the cards in three tiers: silver, gold, and platinum.

A silver card holder can skip all regular customers and proceed to the front of the line, a gold card holder can cut to the front of the line even if a silver card holder is in front, and a platinum card holder can skip everyone and always ensure that he is at the front of the line, except in the case there is another platinum card holder there, in which case he must wait for that person to finish paying or he can just go to another cashier.

A priority scheduler works the same way. In priority scheduling, the scheduler has a task priority, and the process with the higher priority is processed first. If you recall from Chapter 2 on linear data structures, we covered the priority queue.

In case you cannot remember, Figure 11-7 may jog your memory.

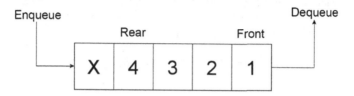

Figure 11-7. A priority queue

In the priority queue, an element with a higher priority will be dequeued before an element with a lower priority. Similarly, in a priority-based scheduling algorithm, processes with a higher priority regardless of execution time will be executed before a lower-priority process.

There are two types of priority-based scheduling algorithms. There is the preemptive priority scheduling and nonpreemptive priority scheduling. In preemptive priority scheduling, when a process with a higher priority level is in the ready-state queue, even if another process is executing, the execution is stopped, and the higher-priority process can execute. Bringing this to our supermarket scenario, if a silver card holder comes to the front of the line, if the system is preemptive, it means that the cashier will immediately stop paying for the current customer and begin to service the silver card holder.

In nonpreemptive priority scheduling, if a process with a higher priority arrives, the current process will finish execution, and then the next process that will execute will be the high-priority process that arrived. In our supermarket scenario, if the system is nonpreemptive, it means that if the silver card holder is currently paying and a gold card holder arrives, the gold card holder must wait until the silver card holder is finished paying and then begin to be serviced.

Processes in a priority queue system can also suffer from starvation because if a lot of higher-priority processes keep entering into the system, then the lower-priority process will never have a chance to execute.

Round Robin

In round-robin scheduling, each process gets an equal share of CPU execution time. In a round-robin system, each process is allocated a time slice with which they may get to use CPU resources. There is a special running time counter that keeps track of how long each process takes. The time slice allocated for running is sometimes called a *quantum*.

In Figure 11-8 we get an idea of how the scheduler allocates processor time in such a system. Each of the labeled processes (P1, P2, P3, P4, and P5) get an equal allocation of CPU processing time, which in this case is two units.

Figure 11-8. Round-robin time allocation

The round-robin system is preemptive by its very nature. A process can execute, and once it reaches the execution time, it is preempted, and another task can run. Round-robin systems utilize the context switching we discussed earlier to switch between tasks.

The round-robin scheduling algorithm is not usually the best method for some systems. This is because some tasks require more processing time than others; in addition, some tasks may not need as much processing time allocated for their execution. This means that if, for example, the two units of time is too little for three of the five tasks, then we will have a lot of unnecessary context switching that takes place. If the two units of time is too long for some processes, then some processes that need the time will have to wait longer, which wastes system resources.

Though there are limitations to this scheduling algorithm, as a system designer you may find there are applications that can benefit from round-robin scheduling.

Multilevel Queue and Multilevel Feedback Queue

There is another scheduling algorithm known as the *multilevel queue scheduling algorithm*. In this algorithm, when processes are in the ready-state queue, processes may be grouped together into different classes where each class has a different scheduling requirement. Each class is then placed into its own queue, and each queue is assigned a priority. Some classes of processes would be okay with FCSC-based scheduling, while some may need round-robin or a priority-based scheduling system.

There is also the multilevel feedback queue scheduling algorithm that fixes some of the problems with multilevel queue scheduling. In multilevel queue scheduling, processes are assigned to a particular queue with a priority level by the very mechanics of the design of the algorithm. This algorithm uses feedback to assign processes to different queues in the multilevel queue system based on their required execution needs.

Conclusion

In this chapter, we discussed operating systems, both general-purpose and real-time operating systems. We also discussed interrupts and finite-state machines. We also looked at several popular scheduling algorithms. In the next chapter, we will briefly look at methods of laying out, designing, and implementing your own algorithms.

Algorithm Planning and Design

We have finally reached the end of our journey. In this chapter, we will look at methods of planning and designing your own algorithms. As a business manager, student, or someone who may have to work with programmers or code, you need a way to communicate your thoughts and ideas to the people who will be implementing the design. You can design algorithms for your own use or for use by a programmer. What I will do in this chapter is to teach you the process and tools you can use for planning and designing your own algorithms that can then be implemented by you or someone else.

The Need for Proper Planning and Design

Before you begin to do anything, you must have a proper plan in place that can be executed. Buildings, cars, bridges, and airplanes all had system designers who envisioned what the final product would look like long before the first cement was poured or the first joint was welded.

© Armstrong Subero 2020
A. Subero, *Codeless Data Structures and Algorithms*,
https://doi.org/10.1007/978-1-4842-5725-8_12

In the realm of computer science, however, it is commonplace that many "coders" will open a terminal and start typing code. The result is that you wind up a program that has poor logical structure and is troublesome for anyone trying to read your program to understand. You see people fail to realize that the act of writing code (*coding*) is not the same as programming and writing proper programs.

To fix this problem, we will approach our algorithm design like a computer scientist, not like a coder. By having a proper method detailing how your algorithm will be implemented, you will be able to design algorithms that are modular, meaning that they will be able to be implemented in any programming language and be used across a wide variety of programs. Take the algorithms and data structures presented in this book; they can be implemented and used in any programming language and adapted to a multitude of paradigms and applications.

Another reason to plan and design your algorithms before implementation is that the methods presented in this chapter are very abstract. Abstraction is the key to realizing the solution to complex problems. This is why in programming implementations come and go, but many of the theories of computation (Turing machines, FSMs, time complexity) have existed for what may seem like eons in the world of modern technology and will be abound for the foreseeable future. These concepts are so powerful and abstract that even though many implementations have come and gone, they have remained the same.

The Three Stages of Algorithms

Let's take some time out to discuss what algorithms are really doing. Algorithms and basically all other computer programs have three stages. They take an input, do some processing, and then produce an output.

During the input stage, the data or information to be processed is loaded into the memory of the machine. The machine will gather this information from a hardware mechanism (an input device, sensor), or it can all come from software. Regardless of where it comes from, before entry into an algorithm, the data must be present in a form that can be processed by the algorithm.

Next, we have the processing stage. In this stage, the data that was input will be actually worked upon. We will perform calculations using the data that was input or maybe do some alteration or manipulation of the data itself (think searching and sorting). At the core of all processing is the CPU. The entire design of your machine and all supporting hardware and software components are to enable efficient operation of the CPU that does the data processing.

The final stage is to produce an output. After your algorithm does processing on the data, it ultimately sends it to some component that uses it. This may be a hardware device like a screen, transducer, or actuator. However, it may also be a software component, and the output may be utilized by another method or function that is running its own algorithm, and your algorithm may just be a cog in a large complex program.

Keep these three stages in mind as you move forward learning how to plan and design your own algorithms.

Flowcharts

The first method we will look at to design our own algorithms is the use of flowcharts. A flowchart is a diagram that we use to represent a process. Flowcharts will allow you to design your own algorithms without having to learn any programming language. Flowcharts are the preferred method of designing algorithms for people learning because they provide a simple pictorial representation of the algorithm you are designing.

Utilizing the power of flowcharts, we can iterate very quickly on our algorithm design and easily spot any logical flaws. Flowcharts also have their program flow in such a way that any modern programming language will be able to use the logic you designed.

In the next few sections, we will look at flowcharts as tools for someone designing an algorithm.

Flowchart Symbols

Before we can use flowcharts, we must become acquainted with their symbols. The flowchart symbols are simple geometric symbols that all have different meanings. In Figure 12-1, we see a list of the flowchart symbols that we will use to realize our algorithm designs.

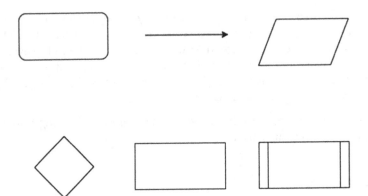

Figure 12-1. Flowchart symbols

Keep in mind that this flowchart list is not exhaustive; however, it is enough of a subset that you can realize almost any algorithm you can fathom and is easily transferable to any modern programming language.

Arrows

If you look at a flowchart, you will observe that there are arrows connecting everything together. See Figure 12-2.

———————————▶

Figure 12-2. Arrow symbol

The arrows within a flowchart will guide someone looking at the program to be able to identify the flow of your chart logic. Arrows are a useful tool that is crucial to any flowchart design.

Terminator

The second flowchart symbol we will look at is the terminator symbol. Figure 12-3 shows use the terminator symbol.

Figure 12-3. The terminator symbol

The terminator symbol is used to indicate the start (entry) and end (termination) points within your algorithm. Your algorithm will have one starting point; however, it is not uncommon, depending on your logic flow, to have several endpoints.

It is easy to identify whether the terminator is a start point or an end point in your algorithm. Good algorithm design dictates that you place the words *Start* or *End* within the shape. Figure 12-4 shows what a start will look like, and Figure 12-5 shows an end.

Start

Figure 12-4. Start terminator

End

Figure 12-5. End terminator

Besides arrows, start and end terminators are two flowchart symbols you are sure to find in your program.

Input and Output

As was previously stated, algorithms have three big stages, which are input, processing, and output. In Figure 12-6 we see the geometric parallelogram shape that is used to indicate an input or an output.

Figure 12-6. The input and output symbol

You usually put you input or output statement in text within the input and output symbol. Figure 12-7 shows what a typical input statement looks like, and Figure 12-8 shows a typical output statement.

Figure 12-7. Input statement

Figure 12-8. Output statement

Process

The process symbol is the workhorse of flowcharting. The process symbol is used to represent basic tasks or actions in your process. Figure 12-9 shows the process symbol.

Figure 12-9. Process symbol

The process symbol is one you will see often in a flowchart because throughout your entire algorithm you will have to take action steps.

Decision

Many times in your algorithms you will need to make a yes/no or true/false decision and then do something else based on that result. The decision symbol, as shown in Figure 12-10, is used to represent this decision.

Figure 12-10. Decision symbol

Once the flowchart determines an answer to the decision, it will take the branch of execution related to that decision.

Predefined Process

If you write every single symbol in a large algorithm in your flowchart, then it may become difficult to follow directly. For that reason, it is sometimes necessary to separate your algorithm into different modules that contain a series of process steps.

The symbol in Figure 12-11 is meant to indicate this.

Figure 12-11. Predefined process

Program Structures

When designing your algorithms, you will notice that some structures repeat quite often. These structures are so common and useful that they have been named. The structures we will look at are the sequence, if-then, if-then-else, while, do-while, and switch.

Each of these structures will be briefly explained in the upcoming sections. The sequence structure falls into a class by itself and is known as a *sequence structure*. The if-then, if-then-else, and switch structures are known as the *selection* or *decision structures*. The while and do-while structures are known as the *loop structures*.

Sequence

The first common structure you will see is the simple sequence. In the sequence, you will have a start terminator followed by a series of processes and an end terminator, as shown in Figure 12-12.

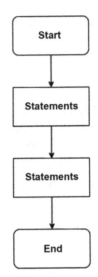

Figure 12-12. The sequence structure

If-Then

Another common structure you can use to design your program is the if-then structure. This structure is used to make a single decision. The if-then structure is shown in Figure 12-13.

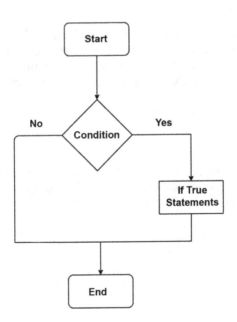

Figure 12-13. The if-then structure

In the if-then structure you have your entry. You make a decision based on the question statement. If your decision is yes, then you will do some action in the process. If the decision is no, then you will exit the algorithm.

If-Then-Else

The if-then-else structure builds on the if-then structure. Like the if-then structure, we test a condition, and if that condition evaluates to true, we take the Yes branch, execute those statements, and then terminate. See Figure 12-14.

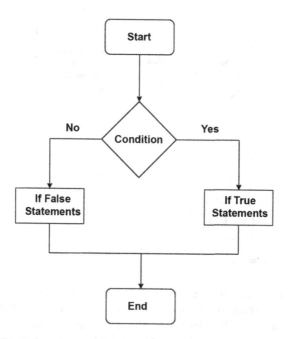

Figure 12-14. The if-then-else

There is another facet to this structure, though. If the decision evaluates to false, instead of just exiting the structure, we take the No branch and then execute the false statements. So, if the condition is true, then execute some statements; otherwise, we execute some other statement.

While Loop

In Figure 12-15 we are presented with another structure, the while loop structure.

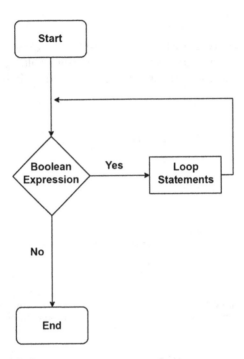

Figure 12-15. The while loop

The while loop is a common structure you will encounter. In the while loop, while the test Boolean expression continues to be true, the loop will keep executing indefinitely. Therefore, for this structure, some exit condition is needed to prevent what is known as an *infinite loop* from occurring where the program runs indefinitely.

Do-While Loop

The do-while loop shown in Figure 12-16 is similar to the while loop.

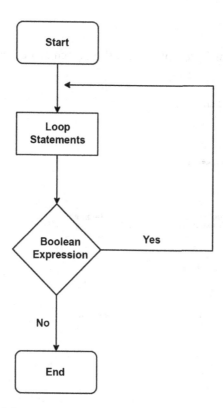

Figure 12-16. do-while loop

This do-while loop, like the while loop, will execute a block of statements at least once. The difference between these two looping structures is that the while loop will execute the statements inside its process only if the Boolean expression is true. In the do-while loop, the process block is run at least once; then it will continue continuously executing the block once the Boolean expression is true in the same manner as the while loop.

Switch

In Figure 12-17 we are presented with the switch structure.

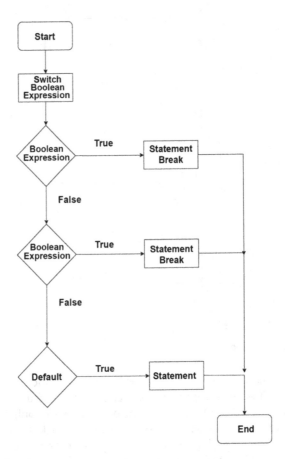

Figure 12-17. Switch case

The switch case structure allows us to execute differing blocks of processes or to branch based on the value of the switch expression. If the switch control statement does not match any of the available cases, then there is usually a default case with default statements that we want to execute.

Example Algorithm Linear Search

Let's put everything we learned so far into the linear search algorithm, as shown in Figure 12-18.

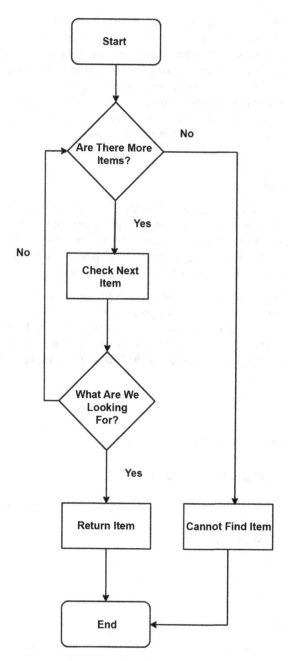

Figure 12-18. Linear search

The linear search algorithm, as you can follow from the flowchart, works as follows. We start with a list of items and ask, "Are there any more items we can work with?" If there are items for us to work with, we check the next item. If we found what we are looking for, we return the item, and the algorithm exits. If we did not find what we are looking for, then we return to the decision of determining whether there are more items. This continues until the "Are there more items?" decision returns false; then we take the No branch, which means that we cannot find the item.

Pseudocode

In this chapter, we learned the building blocks of algorithms using a pictorial representation of our logic. However, there is another way we can use to describe our logic. We can use pseudocode (do you remember our little talk on *pseudo* in Chapter 10?). Programmers favor pseudocode as it gives them a way to use English to describe and model their logic.

Pseudocode is similar to code; however, it is written in such a way that it does not use the syntax of any one language and thus can be modeled in any language. In some implementation languages such as the Python programming language, pseudocode statements can easily be translated into statements of that programming language. Figure 12-19 shows an example of what pseudocode looks like.

```
START

    INPUT FIRSTNUMBER

    INPUT SECONDNUMBER

    SET RESULT = FIRSTNUMBER + SECONDNUMBER

    DISPLAY RESULT

END
```

Figure 12-19. Pseudocode

The pseudocode in Figure 12-19 reads two numbers from the user: firstNumber and secondNumber. We then store the result of their sum in another number called *result* and display it to the user.

Conclusion

In this chapter, we looked at various algorithm design techniques using flow-charts and also briefly discussed pseudocode. With the conclusion of this chapter, we have reached the end of the book. Congratulations! Remember, your journey doesn't have to stop here; you can look at Appendix A to feed your hunger for more knowledge.

Good luck!

Going Further

If you have completed the book, congratulations again! There is a chance you will want some resources to take your studies further. In this appendix, I provide resources to help you succeed.

- First, you will want to check out Project Euler (https://projecteuler.net/). It's the best resource to improve your understanding of data structures and algorithms. Project Euler will help you practice solving different algorithmic problems while learning valuable skills in mathematics.

- Another good resource that you can use to develop your skills is Code Wars (https://www.codewars.com/). You'll use a language of your choice to solve lots of challenges and problems.

- Finally, take a look at FreeCodeCamp (https://www.freecodecamp.org/). While not solely focused on algorithms and data structures, it has a challenging section on algorithms and data structures that will enhance your understanding. As a bonus, you will learn some of the latest web development trends, all for free.

It is my hope that these resources will increase your understanding of data structures and algorithms and help you on your computer science journey.

Good luck going forward!

© Armstrong Subero 2020
A. Subero, *Codeless Data Structures and Algorithms*,
https://doi.org/10.1007/978-1-4842-5725-8

I

Index

© Armstrong Subero 2020
A. Subero, *Codeless Data Structures and Algorithms*,
https://doi.org/10.1007/978-1-4842-5725-8

Printed in the United States
By Bookmasters